AF796162

Osmotically Driven Membrane Processes

Edited by Muharrem Ince and Olcay Kaplan Ince

Published in London, United Kingdom

IntechOpen

Supporting open minds since 2005

Osmotically Driven Membrane Processes
http://dx.doi.org/10.5772/intechopen.91575
Edited by Muharrem Ince and Olcay Kaplan Ince

Contributors
Deniz Şahin, Stavroula Kappa, Simos Malamis, Azzam A. Abuhabib, Rana H. H. Idais, Sofiah Hamzah, Hongbo Du, Raghava R Kommalapati, Layla Ogletree, Muharrem Ince, Olcay Kaplan Ince

© The Editor(s) and the Author(s) 2022
The rights of the editor(s) and the author(s) have been asserted in accordance with the Copyright, Designs and Patents Act 1988. All rights to the book as a whole are reserved by INTECHOPEN LIMITED. The book as a whole (compilation) cannot be reproduced, distributed or used for commercial or non-commercial purposes without INTECHOPEN LIMITED's written permission. Enquiries concerning the use of the book should be directed to INTECHOPEN LIMITED rights and permissions department (permissions@intechopen.com).
Violations are liable to prosecution under the governing Copyright Law.

Individual chapters of this publication are distributed under the terms of the Creative Commons Attribution 3.0 Unported License which permits commercial use, distribution and reproduction of the individual chapters, provided the original author(s) and source publication are appropriately acknowledged. If so indicated, certain images may not be included under the Creative Commons license. In such cases users will need to obtain permission from the license holder to reproduce the material. More details and guidelines concerning content reuse and adaptation can be found at http://www.intechopen.com/copyright-policy.html.

Notice
Statements and opinions expressed in the chapters are these of the individual contributors and not necessarily those of the editors or publisher. No responsibility is accepted for the accuracy of information contained in the published chapters. The publisher assumes no responsibility for any damage or injury to persons or property arising out of the use of any materials, instructions, methods or ideas contained in the book.

First published in London, United Kingdom, 2022 by IntechOpen
IntechOpen is the global imprint of INTECHOPEN LIMITED, registered in England and Wales, registration number: 11086078, 5 Princes Gate Court, London, SW7 2QJ, United Kingdom
Printed in Croatia

British Library Cataloguing-in-Publication Data
A catalogue record for this book is available from the British Library

Additional hard and PDF copies can be obtained from orders@intechopen.com

Osmotically Driven Membrane Processes
Edited by Muharrem Ince and Olcay Kaplan Ince
p. cm.
Print ISBN 978-1-83968-911-6
Online ISBN 978-1-83968-912-3
eBook (PDF) ISBN 978-1-83968-913-0

For EU product safety concerns:
IN TECH d.o.o., Prolaz Marije Krucifikse Kozulić 3, 51000 Rijeka, Croatia, info@intechopen.com or visit our website at intechopen.com.

We are IntechOpen, the world's leading publisher of Open Access books

Built by scientists, for scientists

5,700+
Open access books available

139,000+
International authors and editors

175M+
Downloads

156
Countries delivered to

Our authors are among the
Top 1%
most cited scientists

12.2%
Contributors from top 500 universities

BOOK CITATION INDEX
CLARIVATE ANALYTICS
INDEXED

WEB OF SCIENCE™

Selection of our books indexed in the Book Citation Index (BKCI) in Web of Science Core Collection™

Interested in publishing with us?
Contact book.department@intechopen.com

Numbers displayed above are based on latest data collected.
For more information visit www.intechopen.com

Meet the editors

Professor Muharrem Ince received his Ph.D. in Analytical Chemistry from Firat University, Turkey, in 2008. From 2009 to 2012, he worked as a research analytical chemist at Mus Alparslan University, Turkey. He is currently a professor at Munzur University. From 2013 to 2016, he served as head of the Department of Chemical Engineering, Munzur University. He is an editorial board member of several international journals as well as an author and co-author of more than forty papers published in respectable international journals. His expertise is in analytical method development, spectroscopic and chromatographic techniques, environmental sciences, water pollution identification and prevention, food analysis and toxicology, green and sustainable chemistry, nanoscience, and nanotechnology.

Prof. Dr. Olcay Kaplan Ince received a BS from Hacettepe University, Turkey, and a Ph.D. in Analytical Chemistry from Firat University, Turkey, in 2008. She is a research analytical chemist and previous head of the Food Engineering Department, Munzur University, Turkey. She serves as editor-in-chief of the International Journal of Pure and Applied Sciences. She is the author of more than forty papers published in respectable journals. Her research interests include trace and toxic element analysis, analytical chemistry, instrumental analysis, problem-solving in analytical chemistry, food science and chromatography, nanoscience and cytotoxicity, and deep eutectic solvents.

Contents

Preface

This book, *Osmotically Driven Membrane Processes*, provides an overview of membrane systems and separation processes, recent trends in membranes and membrane processes, and advancements in osmotically driven membrane systems. Over three sections it provides a broad overview of membrane processes for researchers from either an engineering science or chemistry background, starting at a more elementary level, incorporating more advanced concepts, and integrating a wide variety of approaches. The book begins with a section discussing the preparation and applications of nanocomposite membranes for water/wastewater treatment. Chapters in this section address topics such as recent advances in measuring and controlling biofouling of seawater, membrane fabrication technologies, and membrane performances. In the second section, recent trends in membranes and membrane processes are discussed. In the last section, developments about advancements in osmotically driven membrane systems are emphasized.

Written by leading experts in their respective areas, this volume is designed for students, chemists, and other professionals or readers interested in membrane preparation technologies and membrane separation processes.

Muharrem Ince and Olcay Kaplan Ince
Munzur University,
Tunceli, Turkey

Section 1

Membrane Systems and Separation Processes

Chapter 1

Preparation and Applications of Nanocomposite Membranes for Water/Wastewater Treatment

Muharrem Ince and Olcay Kaplan Ince

Abstract

Because of scarcity of clean water all over the globe, it is leads to serious challenges to the survival of all living species. Advanced treatment of water/wastewater techniques such as filtration separation and ion exchange separation are necessary for degradable or non-biodegradable detrimental and hazardous wastes removal from water. Membrane technology is of critical importance to solve this vital problem. In membrane technology, nanocomposite membranes (NCMs) are the most preferred in terms of their convenience. These membranes and their constituent materials are eco-friendly, low-cost, and energy-efficient materials. Also they have operational flexibility and feasibility. The current study presents an overview of the progress in NCMs to treat water/wastewater. To prepare NCMs, various used methods are discussed. Also, to improve the mechanical, antibacterial, and adsorption, properties of NCMs have been investigated. The objective of this work was to summarize the removal of toxic wastes from water/wastewater using various NCMs and to emphasize the shortfalls, and future prospective of NCMs technology are highlighted.

Keywords: nanocomposite membranes, water/wastewater, toxic wastes, membrane technology, adsorption properties

1. Introduction

The rapid growth of the world population and industrial activities has caused a significant increase in water consumption. These developments lead to a serious water shortage all around world especially in arid regions. Recently, the main problem affecting humane societies around the globe is the scarcity of water and increasing demand to it [1, 2]. As stated by United Nations' reports, worldwide, about 1.2 billion people live in the region of physical scarcity. While another half a billion people are approaching this condition, about two billion people are facing economic water shortage. In addition to the treatment and reuse of wastewater, desalination is one of the technologies widely applied in the world. During recent decades, more than 100 countries have been using these processes [3, 4]. It is not possible to survive without clean water, unfortunately, based on international standards and various organizations, less than 1% of total water is clean. Rest of the water quantity is contaminated by various human-source pollutants such as agricultural activities, municipal wastewater, and industrial wastes [5, 6]. The major

IntechOpen

water pollutants can be specified as toxic heavy metals, pesticides, dyes, organic acids, halogenated compounds, fertilizers, and microorganisms [7–10]. Because of non-biodegradability and toxicity, among these pollutants, heavy metals are the most hazardous materials for ecosystem and organism, because these toxic and dangerous metals tend to accumulate in ecosystem especially the food chain and the living organism. In addition, the polluted water intake leads to various health problems, such as organ damage, skin irritation, cancer, rupture of nasal septum, diarrhea, appetite loss, abdominal pain, and headache [11–14]. For the reasons stated above, and especially in order to provide clean water to all living creatures to survive in a healthy life, these pollutants must be removed. Various membrane technologies have recently been used for removing these contaminants from water/wastewater. Among these technologies and applications, those of the greatest interest to researchers are listed below:

- Microfiltration (**MF**; range from 0.05 μm to 1.0 μm),
- Ultrafiltration (**UF**; range from 0.005 μm to 0.5 μm),
- Nanofiltration (**NF**; range from 0.0005 μm to 0.01 μm),
- Reverse osmosis (**RO**; range from 0.0001 μm to 0.001 μm),
- Forward osmosis (**FO**),
- Membrane distillation,
- Pressure retarded osmosis (**PRO**),
- Membrane bioreactor (**MBR**),
- Pervaporation (**PP**), and
- Separation using liquid membranes [10].

In the last decade, various water/wastewater purification technologies such as NF, FO, and RO have been developed and effectively used [15]. It is inevitable that membrane-based processes will play an increasingly important role in water/wastewater treatment. These processes are expected to take a key role in solving many problems by developing further in a short period of time due to some advantages such as requiring less energy, ease of use, and making them easily modular [16]. Among the membrane technologies performed in the wastewater treatment, especially application of NF, FO, and NF processes will be increased in the near future [1]. Polymers are widely preferred materials in water/wastewater treatment, despite some disadvantages such as relatively high-energy consumption, permeability, short lifetime, relatively consumption of high energy, and low resistance to fouling. It is vital to develop low-energy, cost-effective, and functional membranes for contaminants removal from water/wastewater. In particular, the inclusion of nano-sized materials in the polymer matrix has made a significant progress in overcoming the challenges of water treatment of polymeric membranes developed and synthesized. Studies conducted in the last few years, especially nano-sized structures such as carbon nanotubes (CNTs), graphene, zeolites, silica, zinc, iron oxide, and other metal oxides, have been added to the polymer matrix and tested [17]. Supported by various nanostructures, NCMs have been used effectively in

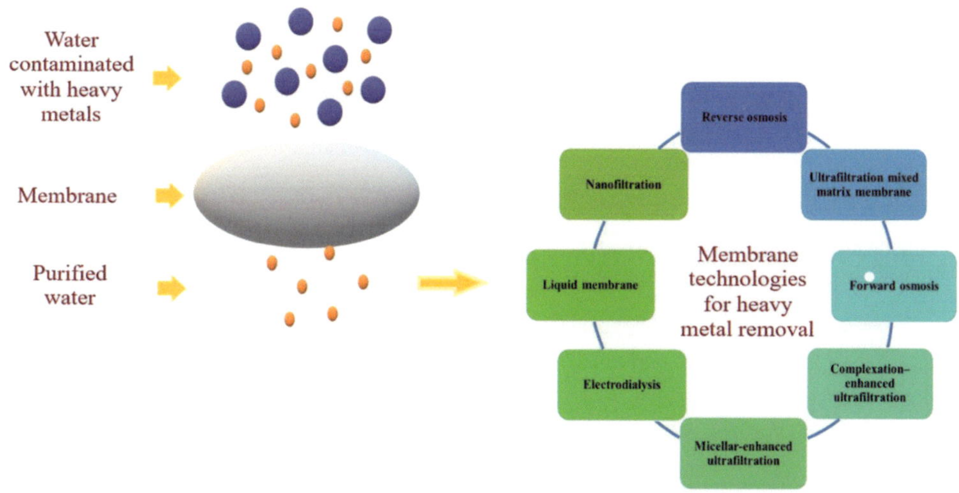

Figure 1.
A schematic illustration of NCMs for heavy metals removal from wastewater.

many applications including liquid-solid, liquid-liquid, and gas-gas separations. The PCMs have attracted great attention for water/wastewater cleaning because of overcoming trade-off between permeability and solute rejection along with fouling reduction property. Also, for water/wastewater treatment process, they are known as high-performance membrane [18]. As a result, it can be clearly stated that although there are some difficulties in industrial applications, nanomaterials offer outstanding benefits. For example, modification of the NCMs' surface provides a great advantage in water treatment applications as it significantly changes its efficiency, such as pore size and hydrophilicity [18]. A brief schematic summary of NCM processes to treat water/wastewater is presented in **Figure 1**.

In order for designing membranes for water/wastewater treatment, various natural and synthetic polymer types, including chitosan, cellulose acetate, polystyrene, polyamide, have been preferred [10, 19, 20]. Barriers including low contaminant removal, low chemical stability under pH change, biological fouling, loss of mechanical strength, and hydrophobicity prevent the widespread application of the polymeric membrane. The advantages and disadvantages of polymers used in NCMs are given in **Table 1** [10].

For water/wastewater treatment, nanotechnology has brought a great revolution. During the formation of polymeric nanocomposite membranes (PNCMs) process, when nanoscale entities such as nanoparticles and nanofibers add to PNCMs; it gives them unique properties. In the water/wastewater treatment processes, because of some outstanding properties such as permeability, mechanical and chemical stability, superior flexibility, less installation space requirement along with selectivity to chemical species, and high removal capability, PNCMs have become an ideal choice. The significance of PNCs for water/wastewater treatment can be tracked by the continuous rise in publications, also. Using PNCMs for treatment of water/wastewater is an energy-efficient eco-friendly and technology besides low-cost. Moreover, PNCMs technology can be feasibly combined with various processes [10, 21]. For example, the inclusion of metallic and metallic oxide nanoparticles in the polymer matrix has added antifouling properties to the membranes, as well as increased thermal and mechanical stability. Because of their low cost, they are often added in a small quantity of nanofibers into the polymer matrix. When the nanofibers are added to polymer matrix, in addition to the mechanical strength and thermal stability of polymer-based nanocomposites, its flexibility also increases [22–25]. Nanoparticles

Polymer	Advantages	Disadvantages
Chitosan	• Chemical stability, • Fabrication flexibility, • Chelating properties, • Insoluble in water, • Film forming ability	• Failure in acidic pH range
Polyvinyl alcohol	• Good mechanical properties, • Low thermal properties and strength for few applications, • Fabrication flexibility, • Chemical resistance, • Limited biological performance, • Film forming ability	• Dissolve in aqueous solutions
Cellulose acetate	• Hydrophilicity, • Fabrication flexibility, • Low cost	• Poor chemical resistance, • Poor thermal resistance, • Poor chlorine resistance
Polyamide	• Perfect thermal stability, • High mechanical properties, • Wide pH range	• Poor chlorine resistance
Polystyrene polyethersulfone	• High mechanical properties, • Thermal resistance, • Wide pH range, • Chlorine resistance, • Fabrication flexibility	• Less operating pressure limit, • Hydrophobicity

Table 1.
Advantages and disadvantages of some commonly used polymers in NCMs [10].

composed of metal or metal oxides, which contain Ag, Cu, TiO_2, and Fe_2O_3, are main examples of nanoparticles [26]. This nanoscale entity class of spherical shape demonstrates some superior properties such as increased lipophilicity and good dispersibility in organic solvent along with chemical stability. Another important class of nanoscales are carbon nanotubes (CNTs), nano-diamonds, and graphene oxides (GO). The GO is a 2D carbon-based nanomaterial, and they contain many oxygenated functionalities such as carbonyl or hydroxyl groups in their interconnected carbon layers. The CNTs with 1D carbon-based tubular layers have often been used in wastewater treatment process. Because of some vital properties including hydrophobic surfaces and low surface energy besides spherical nanoscale such as Cu, TiO_2, ZnO entities are often used as disinfectants or antimicrobial agents [26]. Very small spherical nanoscale entities such as nanofibers, nanoplatelets, or polymers can be grown at the surface of the substrate to obtain functional nanoscale structures for applications such as catalysis. Various nanostructures obtained in this way are used for water purification as environmentally friendly, cost-effective, and quality products [10, 27]. A schematic illustration of pressure-driven NCMs for water/wastewater treatment is presented in **Figure 2**.

This study provides an overview of the applications of these current technologies in water/wastewater treatment for heavy metal removal, focusing on the latest technological developments in this field, as well as techniques for preparing NCMs. The properties and performance of PNCMs will be discussed considering different

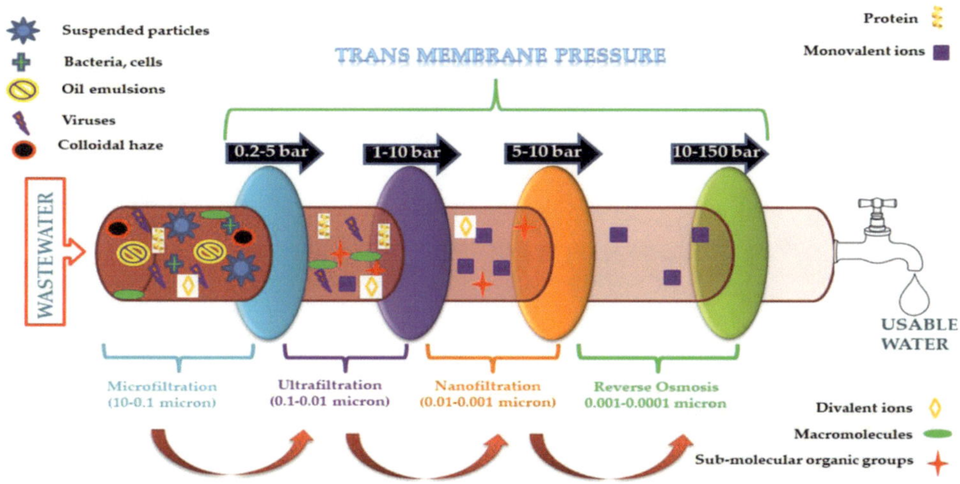

Figure 2.
A schematic pressure-driven membrane nanocomposite illustration for wastewater treatment.

polymers and nanoscale entities. It is expected that this study would arouse curiosity and interesting for the development and application of functional NCMs to treatment water/wastewater. Moreover, preparation and application of the NCMs and future prospectives will be discussed.

2. Preparation techniques of PNCMs

Two main factors such as the nature of the polymer and the final required membrane structure play a key role in determining the type of membrane preparation method. Several different techniques may be used for fabricating of polymeric membranes. However, in general, some techniques are often preferred among them, and these are summarized below:

- Interfacial polymerization technique
- Phase inversion technique
- Blending technique
- Electrospinning technique [28].

Unfortunately, current membrane preparation techniques are not suitable for industrial-scale use. Therefore, efforts to develop and promote preparation techniques to overcome the limitations of existing strategies for membrane preparation should be strongly supported (**Figure 3**).

2.1 Interfacial polymerization technique

The technique called interfacial polymerization (IP) is often preferred for PNCMs production, and it is the most considerable technique for commercially fabricating various essential and important membranes including nanofiltration (NF) and thin-film composite (TFC) besides reverse osmosis (RO). For RO applications, the development and use of interfacial polymerized TFC

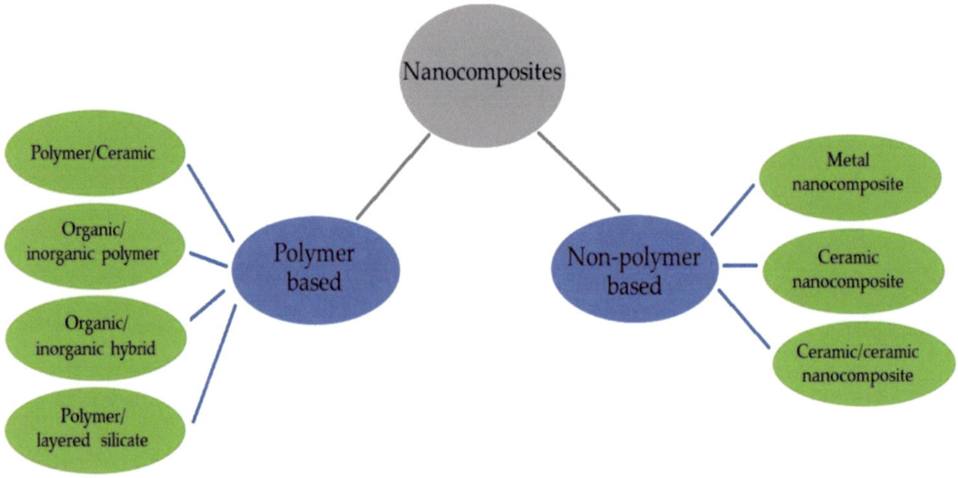

Figure 3.
A schematic illustration of nanocomposite materials.

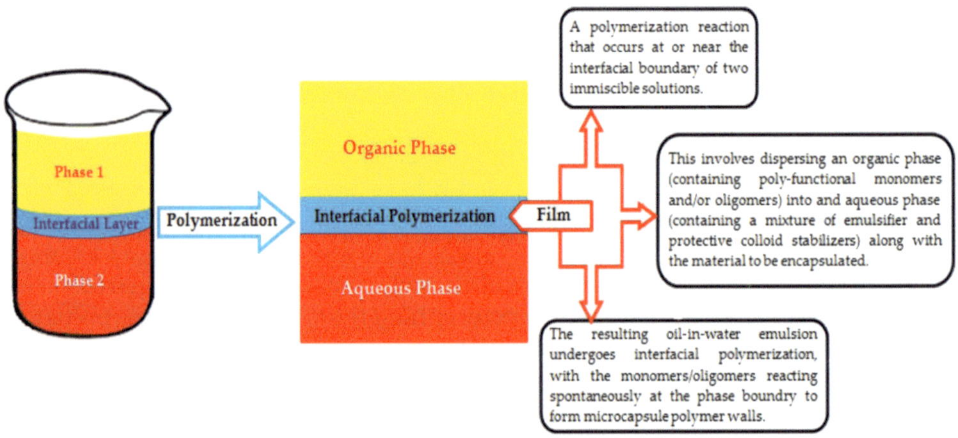

Figure 4.
A schematic illustration of interfacial polymerization technique for preparation of PNCMs.

membranes were an important milestone in the performance of the membranes [29]. Various types of TFC membrane have been fabricated using IP technique because of noticeably superior properties such as microporous substrate layer and independent optimization of the skin layer. In addition to TFC membranes, RO and NF membranes, which have many uses, were also produced using this useful technique. Barrier membrane layer composition and structural morphology are affected with many factors including monomer concentration, solvent type, reaction time, and subsequent treatment [30, 31]. In the highlighted technique, the interaction of two different monomers is as schematized in **Figure 4**. Before the polymerization process begins, nano-sized structures are incorporated and polymerization occurs at the interface between the two phases. The distinctive layer produced on the substrate in IP has very less thickness and due to these properties, possesses superior membrane flux. Also, for polymerization, suitable monomers selection can produce selective polymer layers, resistant to chemicals, good thermal stability, and better durability [10, 28].

2.2 Phase inversion technique

For membrane preparation, the most usable method is phase inversion (PI) technique. For different applications, various kinds of morphologies can be obtained when using IP technique [32]. To design membranes, various polymers can be used effectively because the PI method is cost-effective, time-efficient, facile, flexible in use, and feasible to scale-up; therefore, PI technique is used for the manufacture of lab membranes and commercial [33]. All PI membranes are based on polymer precipitation in homogeneous casting solution. The polymer solution precipitation is governed by PI process kinetics and thermodynamics. Therefore, this process affects the prepared membranes final morphology. The precipitation takes place through a demixing process. In this mixing process, the polymer solution is converted from a liquid to a porous solid due to the exchange between solvent and non-solvent [34]. In other words, the PI process is a uniform polymer and a mixing process in which the solution of nanoscale structures is converted into a solid phase in a controlled manner. The PI technique is based on the change in solution stability of the dissolved polymer and nanoscale entity caused by temperature instability, mass change between coagulant bath/non-solvents, and solvent evaporation. The PI technique is frequently selected for preparation of asymmetric PNCMs with thin and dense layers. For membrane preparation (**Figure 5**), some parameters including solvent, non-solvent, choice of composition of polymer solutions, coagulation bath composition, and film casting conditions are the key parameters. As result, PI technique provides the advantage of large morphological differences by altering process parameters that are mentioned before [35, 36].

2.3 Blending technique

Blending technique (BT) is based on the direct mixing of nanoscale entities such as metal oxide or metal nanoparticles and polymer. Also, to form PNCMs, the BT is the easiest technique. Such mixing/mixing can be accomplished in two ways, solution blending technique and melt blending technique.

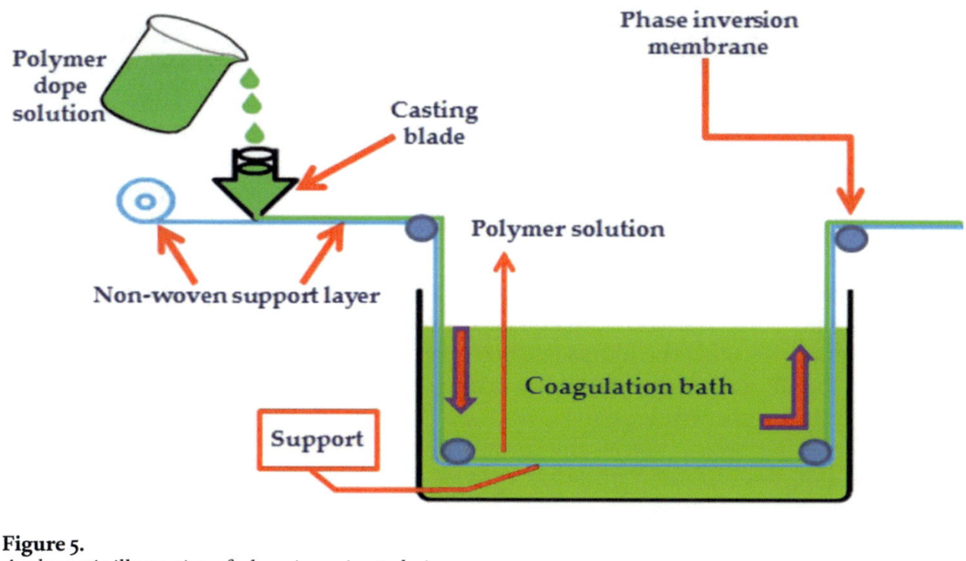

Figure 5.
A schematic illustration of phase inversion technique.

2.3.1 Solution blending technique (SBT)

In this technique, whole components such as polymer and nano-dimensional structures are disseminated in an appropriate common solvent. It is supported in the polymer matrix by adding nano-sized structures to the mentioned solvent. In this step, first the surface-modified nanoscale entities are dispersed in the solvent by means of ultrasonic waves. The mixture is then added to the polymer solution to obtain a homogeneous dispersion. The polymer chains remained intact and formed NCM upon solvent evaporation. Also, to select a convenient solvent for proper nanoscale entities and polymer mixing, it should not be forgotten that it is essential. For water-soluble polymers, the SBT is more useful. However, using organic solvents that are toxic and expensive is a main problem in application of SBT [37]. If NCM is to be obtained from polymers insoluble in low-boiling-point solvents, SBT is not preferred [38].

2.3.2 Melt blending technique (MBT)

NCMs are formed from the polymer melt, after the nanoscale structures and polymers are injected into the extruder and melted at high temperature during intensive mixing. For providing polymer chains mobility, thermal energy is used. Generally, the MBTI is preferred because of its environment-friendly nature along with higher effectiveness. At a large scale, some parameters including use of high temperature and setup of processing may lead to limitations to their use [39]. During blending, the extruder configuration and screw affect the nanoscale entities dispersion quality, also [38, 40]. For preparation of NCMs, both SBT and MBT are simple and frequently used techniques. Generally, BTs are feasible to operate and appropriate for all nanoscale entities types.

2.4 Electrospinning technique

Electrospinning technique (ET) that is simple and effective method is usable for producing fibrous membranes. Because of a number of uses including filtration and desalination, the ET is relatively new. It is a preferred technique for fabricating particularly porous membranes. The ET is often preferred for membrane preparation due to some superior features including excellent interconnectivity besides relatively homogeneous pores distribution. Also, because of nanofibers' large surface area, these membranes functionalization are easier. The ET is a durable technique with very good control over the membrane structure. On the other hand, dense membranes are not produced using ET, because these membranes are required for a diffusion processes, for example, NF and RO [41]. In ET, an application of high potential using a voltage source is made between polymer solution droplet and grounded collector. When electrostatic potential is raised to an adequate level, it overcomes the droplet's surface tension and forms a charged liquid jet. The solution of polymer contains the nanoscale entities and dissolved polymer. The fiber-containing membranes are perfect in that the fiber and morphology of the aspect ratios of the nano/microfibers can be controlled by variable parameters such as the applied electrical potential level, the flow rate of the membrane solution, and the membrane solution viscosity [10, 42, 43].

Finally, because of the forces present between them, jet leaves the tip. During thinning of the polymer jet, solvent phase evaporates and nanofibers are formed. Then, nanofibers are collected on the collector. Owing to rheological properties, jet instabilities arising of polymer melt are important in the shaping of fibrous membrane. By controlling the parameters and operating conditions, the fibrous membranes properties including morphology, porosity, aspect ratio, pore size distribution, and hydrophobicity can be regulated. Electrospun fibrous membranes are a preferable choice in applications filtration, because morphology, fiber shape, and size can be precisely controlled. It has been reported in many studies that nanostructured morphology and fiber diameter can be significantly affected by applied potential strength, solution feed rates, ionic salt addition, and polymer solution viscosity. Depending on the chosen polymer and its molecular weight, the minimum viscosity is decided [28, 42, 44].

3. Antibacterial, mechanical, and adsorption properties of NCMs

In the processing and application of NCMs for the water/wastewater treatment, biofouling is one of the main drawbacks of membrane technology. Biologically sourced membrane contamination leads to clogging of the pores and thus to a serious decrease in performance. Moreover, biofouling increases the maintenance and operational cost of membranes. It decreases the membrane average lifetime, also. Microbial increase and biofilm formation are the main problems that increase the flow in the membranes and consequently require more energy. To overcome these problems, it is vital to prepare NCMS with antimicrobial activity. Preparation of NCMs with antimicrobial activity both increases the efficiency of the membranes and saves time by shortening the application time. Recently, many researchers have focused on using polymers with biocidal materials in designing NCMs with antibacterial properties. Metal oxides such as Ni_2O_3 [45], TiO_2 [46], and ZnO Al_2O_3 [47] are frequently preferred because of their biocidal properties, that is, they directly target bacteria. For this reason, to reduce biofouling, metal oxides are commonly used to design the antibacterial NCMs. For example, owing to the outstanding biocidal properties of Ag, it is one of the most studied nanomaterials to create antimicrobial activity. Other nanostructures such as titanium, chromium, and copper are also metals that are highly preferred in antimicrobial NCMs production [18]. The optimal concentrations of most metal oxide nanoparticles used to destroy bacterial cells have no toxic effects on human health, which has also fueled interest in the use of these materials. The PNCM antimicrobial effectiveness is based on the electrostatic interaction between the membrane and bacteria. Commonly, the nanoscale structure found in PNCM contains a positive charge that attracts the negatively charged bacterial cell on their surface. This electrostatic interaction breaks the structural integrity of the bacteria and leads to the bacteria death [48].

The NCMs, during water/wastewater treatment, must possess good mechanical features including toughness, to endure the pressure. Because to define processability and stability besides end use of NCMs, enough mechanical strength is essential. The interaction between nanoscale entities and polymer components is vital in NCMs that impart mechanical properties. Nanoscale entity uniformity, size, and volume fraction affect mechanical properties. The good and homogeneous nanoparticle distribution in the polymer matrix restricts the chain movements and thus increases the mechanical strength. As a result, it can be clearly stated that anisotropy is an important property that is also responsible for its mechanical properties [11, 49, 50].

Several technologies are available for removing pollution including organic and inorganic from water/wastewater. The removal techniques such as chemical precipitation, coagulation/flocculation, membrane processes, reverse osmosis, ion-exchange/solvent extraction, biological operations, ultrafiltration, and adsorption have been used. Other techniques including precipitation and ion-exchange other than adsorption are not preferred because of the production of various secondary pollutants and their high operating efficiency. Apart from these techniques, the adsorption technique has come to the fore and has been accepted due to some advantages such as simple, efficient, and cost-effective. Adsorption that is most effective techniques is often preferred to remove heavy metals due to flexibility in design and operation. This technique contains a surface phenomenon where pollutants are deposited over the adsorbent surface. Ekstra energy, excess water, or additional chemicals are not used in adsorption process [26]. For the aforementioned reasons, the adsorption technique has become a unique phenomenon in removing contaminants from water/wastewater. Adsorbate and adsorbent surface interaction called physisorption or chemisorption occurs in adsorption process. The pollutants especially heavy metals may interact with the adsorbent surface with various forces including electrostatic interactions, van der Waals, or hydrogen bonding [26]. Functional groups included in PNCMs take part in pollutants and heavy metal ions removal by adsorption and can be regenerated by desorption process [51]. In recent years, various materials including nanoparticles [51] and beads [52] as adsorbent forms have been developed and used. In the last few years, many composite materials such as graphene oxide have widely used to remove heavy metal contaminants as novel adsorbents for the adsorption.

4. New trends for removing hazardous metals from water/wastewater using advanced membrane technologies

Detrimental heavy metals such as As and Ni are the biggest and most important pollutants for ecosystem. These toxic and carcinogenic pollutants can be discharged into the water sources in almost all walks of numerous industrial activities. They have damaged the environment and human health in many aspects. Since these metals, which are harmful and destructive, can enter the human body at more than the allowed concentration and accumulate in our tissues, they cause various harmful health problems. Since metals that are toxic effect are used in many fields of industry, without discharge of their release to the environment is also increasing. Toxic materials especially heavy metals, which spread to the environment and do not degrade, reach people especially through the food chain and water [53–55].

As practical and environmental approach for treating wastewater, separation technologies using membranes have been known as worldwide one of the best technology [56]. Membranes used for this purpose can be divided into two basic classes: inorganic membrane and polymeric membrane (**Figure 6**). There are four types of membranes, based on pore sizes, which are MF, UF, NF, and RO. It should not be forgotten that, during the water/wastewater treatment, heavy metals ions such as Ni^{2+}, Cd^{2+}, and Hg^{2+} were tiny, and sometimes they are soluble in which

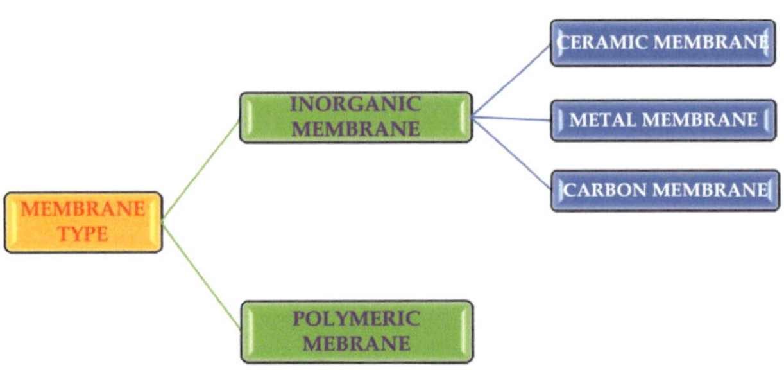

Figure 6.
A schematic illustration of membrane type.

it is necessary to reverse the osmosis membrane's size [57]. Therefore, recently, materials known as hybrid or specifically adsorptive membranes produced by the combination of adsorption and membrane separation processes have been the focus of many researchers [58, 59].

Because of sieving and surface charge effects, both NF and RO can effectively be used for heavy metal ions removal [58]. At the same time, for modification conventional UF and MF membranes to improve the membranes selectivity toward heavy metal ions, various studies have been made. Studies conducted in recent years for heavy metal ions removal from water/wastewater are summarized below.

For selective ion removal from water, Ag-doped multiwalled carbon nanotube (MWCNT)/polyphenylsulfone (PPSU) was prepared as NCM by Shukla et al. Silver-doped MWCNTs prepared and characterized based on specific surface area and distribution of particle size. For characterization, various properties such as porosity, topography, morphology, surface charge, and contact angles were investigated. To examine mentioned properties, several spectroscopic techniques besides transmission electron microscopy were used. It was stated that Ag-MWCNT/PPSU NCM achieves optimal performance and exhibits unique properties. When PPSU membrane is compared with NCMs, it was mentioned that NCMs exhibit significantly improved selective removal of several ions such as Na^+, As^{5+}, and Mg^{2+} ions from aqueous medium. Also, antibacterial activity of Ag-MWCNTs was evaluated using some bacteria such as *Escherichia coli* and is reported that the Ag-MWCNTs exhibited excellent antibacterial activity. Finally, it was emphasized that the applications of developed nanocomposite Ag-MWCN/PPSU membranes, which have antibacterial activity, in removing several metal ions in drinking water applications can be performed successfully [59].

Delavar et al. reported the removal of Cd^{2+} and Cu^{2+} ions using mixed matrix membranes (MMMs) alumina nanoparticles fabricated as UF membranes and incorporated with alumina nanoparticles [60]. The characterization of structural morphology and hydrophilicity of synthesized MMMs was made by using field emission scanning electron microscope (FESEM), water contact angle, and Fourier transform infrared spectroscopy (FTIR) techniques. The alumina and hydrous manganese oxide (HMO) loading affected some properties such as pure water flux, mean pore size, porosity, and water contact angle of the membranes. In the light of this information, the performance of UF membranes for removal of $Cd^{2+}$and Cu^{2+} ions was also investigated. Based on obtained data from UF experiments, when prepared MMM with a high HMO nanoparticles loaded, it was stated that they have had very fast kinetics and demonstrated the highest Cu^{2+} ions and Cd^{2+} ions removal efficiency (97% and 98%, respectively). This study results indicated that HMO nanoparticles can be a good candidate for preparation of MMMs. Also, to remove Cu^{2+} ions and Cd^{2+} ions from polluted water resources, it can be conveniently used [60].

In another study, to remove Cr(VI) ions using UF membranes that contain cellulose acetate, this is incorporated with TiO_2 nanoparticles [61]. In addition, TiO_2 nanoparticles were preferred to increase the affinity of heavy metal ions to the membrane and increase the removal efficiency. Moreover, TiO_2 nanoparticles' presence improved the membranes' antifouling properties because of easily cleaned and regenerated. At pH 3.5, in the presence of aminated TiO_2 nanoparticles, Cr(VI) removal efficiency was achieved as 99.8%. Present in the anions form such as Cr(VI) ions, the protonated amine group on the TiO_2 nanoparticles established electrostatic interaction with the Cr(VI) species. Gebru and Das also reported, after four cycles of washing and regeneration processes, efficiency of removal was only slightly reduced to 96.6% [61].

In another study, for the treatment of water contaminants such as organic fouling agents and toxic heavy metal ions, a] thin film composite (TFC) NF membrane that contains poly(piperazineamide) [poly(PIP) was developed by Bera et al. It has been reported that the synthesized NCM has high performance in anti-organic fouling, anti-biofouling, and removal of multivalent cations. Also, they reported the thin film nanocomposite (TFNC) NF membranes preparation with improved rejection of heavy metals efficacy, anti-biofouling property, and anti-organic fouling properties compared with that of poly(PIP) TFC NF membrane. Using IP technique, FNC NF membranes were prepared and PEI-polyethylene glycol conjugate and then immobilization of Ag-NP. The IP was performed on a polyethersulfone/poly(methyl methacrylate)-co-poly(vinyl pyrollidone)/Ag-NP mixture UF membrane support. The synthesized TFNC membranes exhibited a good performance for several heavy metals as >99% for Pb^{2+}, 91–97% for Cd^{2+}, 90–96% for Co^{2+}, and 95–99% for Cu^{2+} at applied 0.5 MPa pressure. It was mentioned that heavy metal ions rejection effect of the modified NF membranes is attributed to the positive surface charge development [62].

In the study carried out by Deng et al., a novel NCM containing improved physical properties and enhanced metal ions removal efficiency was prepared using ET technique. By reacting MWCNT-COOH with polyethylenimine (PEI), modified MWCNTs were fabricated, which was further embedded within polyacrylonitrile (PAN) nanofibers using ET technique. The MWCNT-PEI and NCM (MWCNT-PEI/PAN) properties such as physical properties, morphology, and structure were characterized using various techniques including TEM, SEM, FTIR besides mechanical test and contact angle measurements. When NCMs compare to plain PAN membrane because of hydrophilicity, higher mechanical strength, high permeation, and filtration efficiency, it is undisputed that the NCMs are clearly superior. Experiments studies revealed that synthesized NCMs such as MWCNT-PEI/PAN exhibited higher adsorption capacity for several heavy metals such as Cu^{2+} and Pb^{2+} ions compared with other NCMs. It was reported that Langmuir isotherm and dynamic adsorption results proved that the synthesized and designed NCMs exhibited improved rejection ability for heavy metal ions with a water flux at 145.8 L m^{-2} h^{-1} under 0.2 bar pressure. It is stated that these new and functional membranes synthesized have promising potential for contaminated water treatment due to their heavy metal removal properties [63]. In another study carried out, to functionalize graphene oxide (GO) nanoparticles using polyaniline (PANI), a polymerization technique was performed by Ghaemi et al. After NCMs were prepared by embedding PANI@GO nanoparticles into matrix of PES membrane, it was characterized by SEM and AFM for measuring various properties including porosity and permeability besides mean pore size. A response surface methodology compatible with central composite experimental design was carried out for membrane Pb^{2+} removal performance from water besides to optimize experimental conditions. Although the NPs addition to membrane matrix reduces the porosity, permeability, and hydrophilic properties, it has been reported that Pb^{2+} removal activity is significantly increased. It has been stated that increasing the pH and increasing the proportion of GO particles up to 25% by weight cause an increase in Pb^{2+} removal from the water and almost all Pb^{2+} ions are successfully removed by the NCMs. Ghaemi et al. examined adsorption mechanism, isotherm model, and the kinetic properties along with reusability performance of NCMs. They reported that Langmuir isotherm and pseudo-first order kinetic offered the most appropriate models for Pb^{2+} removal from water using synthesized NCMs [64].

Gohari et al. developed an UF mixed matrix membranes (MMMs) using polyethersulfone (PES)/hydrous manganese dioxide (HMO) for Pb^{2+} adsorption and removal by varying the weight ratio of PES:HMO in the membrane. The HMO loading effect on the membrane in terms of porosity, hydrophilicity, and pure water flux and adsorption capacity MMMs for Pb^{2+} studied, also. Moreover, prepared

Used membrane	Removed metal	Reference
PAH and PSF/PAH blend membranes	Pb^{2+}, Cd^{2+}	[67]
Cellulose NCMs	Ag^{+}, Cu^{2+}, Fe^{3+}, Fe^{2+}	[68]
FO membranes	Co^{2+}	[69]
CTA-ES membrane	Cs^{+}	[70]
NCMs (PDA/GNPs/PEI)	Zn^{2+}, Ba^{2+}, Ni^{2+}, Cd^{2+}	[71]
Fe-Ag/f-MWCNT/PES	Cr(VI)	[72]
Thin-film nanocomposite FO membrane	Pb^{2+}, Cd^{2+}, Cr^{6+}	[73]
Ultra-thin NCMs (chitosan/GO NCM)	Mn^{2+}	[74]
MWCNT/polysulfone composite membranes	Cr^{6+}, Cd^{2+}	[75]
Ag-doped MWCNT NCMs	As^{5+}, Cr^{6+}, Mg^{2+}	[59]
NF membranes(quaternized polyelectrolyte complex membranes)	Na^{+}, Mg^{2+}, Ca^{2+}, Cu^{+}, Mg^{2+}, Zn^{2+}	[76]
Thin-film inorganic forward osmosis membrane	Cd^{2+}, Pb^{2+}, Cu^{2+} and Zn^{2+}	[77]
Functionalized MWCNTs/PVA nanocomposite films	Zn^{2+}, Ni^{2+}, Mn^{2+}, Cr^{2+}, Cd^{2+}, Pb^{2+}	[78]

Table 2.
Heavy metal removal from water/wastewater by using various NCMs.

membranes properties such as structural morphology using and surface roughness were characterized by using SEM, AFM, and FTIR techniques. It was reported that in spite of pore size of membrane tended to decrease with increasing PES:HMO weight ratio, it has been stated that the water flux of the membrane is not affected. It was found that the Pb^{2+} removal capacity of the MMM prepared with the highest PES:HMO ratio was 204.1 mg g^{-1} and this adsorption capacity was quite promising, when compared with literature. It was observed that Pb^{2+} adsorbed on the membrane can be easily desorbed by using HCl solution. Studies conducted by repeating the adsorption-desorption process proved that this MMM can be used repeatedly [65].

In another study, the synthesis of polymeric membranes based PES and modified by the activated carbon addition and the removal potential of this developed membrane in removing Cu^{2+} ions from the aqueous medium were investigated. It has been reported that after modification of the PES membrane with the activated carbon addition, the retention capacity of Cu2+ ions are significantly increased [66]. Moreover, the synthesis of various membranes and their application to various samples for heavy metal removal are summarized in **Table 2**.

5. Conclusions: suggestions and future perspectives

Recently, for water/wastewater treatment, PNCMs that have perfect antibacterial, mechanical, and adsorption properties and have become a globally known and usable method. Because of these outstanding performances, they managed to attract the attention of academia and industry. It is a variety of functional nanoscale materials and different architectures that allow PNCMs to have some outstanding properties. It has enabled an impressive improvement in the treatment of water/wastewater using PNCMs technology, which is open to this and similar developments. Over the past decade, to treat water/wastewater especially for removing toxic heavy metals, immense progress has been made in developing PNCMs. Membranes exhibit a unique useful behavior due to some of their physicochemical properties including charge density, porosity, hydrophilicity, mechanical and thermal stability, and nanoscale entities addition. Although two important parameters such as the polymer nature and the final required membrane structure determine the type of the method used in the preparation of the membrane, different methods such as IP and PI are the most frequently used methods for PNCMs preparation. The addition of nanomaterials such as nanofibers is of vital importance in improving the mechanical and, in particular, adsorption capability of membranes.

It can be clearly stated that spherical nanomaterials, for example, metal oxide nanoparticles, protect the membrane against biological contamination and provide antibacterial activity. However, it is clear that the methods for the synthesis, development, and distribution of nanoscale materials in the polymer matrix need more research. These research studies should focus especially on the following subjects:

- The aggregates formation is a major problem for nanoscale entities dispersion into the polymer matrix as homogeneous.
- The compatibility of polymer and nanoscale materials plays a key role in the formation of a successful membrane.
- Focus should be placed on possible solutions for the stability of nanoscale entities in the polymer matrix that affect membrane performance.
- Further studies are needed on the functionalization of the surface of nanosized materials or optimization of the manufacturing process, the ability to increase the dispersion of nanoscale entities into polymer matrices. In the near future, it will be possible to optimize the distribution and hence the interaction between nanoscale entities and the polymer matrix.

While studies have shown that nanomaterials have unique properties that can contribute to the development of high-tech and new NCMs with advanced capabilities to treat water/wastewater, optimizing the durability of nanoscale assets and the loading concentration in NCMs is key to achieving the best performance. On the other hand, recently, for removing or reducing the heavy metals amount in water/wastewater bodies, extensive efforts have been made. Various methods have been applied, each of which has advantages and disadvantages. For removing of toxic metallic ions, membrane technology including UF, NF, RO, and FO membranes exhibits huge potential as it offers different separation mechanisms and a wide range of membrane properties. One of the best examples of this is that the adsorptive UF membrane shows a significant improvement in membrane morphology where the increase in water permeability is achieved. Even at low membrane pressure, mentioned membrane is convenient to treat low metal concentrations by enabling the complete filtration-adsorption metal ions removal. The NF, RO, and FO membranes have proven effective to remove metal ions from water/wastewater. In addition, the NF process has been reported to be efficient and effective even at an industrial level. Because it requires high energy consumption and is therefore costly, the RO method is preferred only to treat water resources that must meet drinking water standards. For these reasons, to reduce costs and expand usage, more research is still needed for RO. If the respective limitations can be overcome, it is highly likely that membrane technology will become a serious alternative method to remove heavy metal ions in the future. In summary, adsorptive membranes have a wide range of applications including wastewater treatment. Advances in the development and manufacture of adsorptive membranes are increasing day by day. Besides adsorptive removal of contaminants from the water/wastewater has also technologically advanced, it has played an important role with the development of adsorptive membrane technologies. Thus, multi-stage pollutant removal processes, which were previously applied, can now be carried out in a single-stage pollutant removal process. As a result, the contribution of properties such as pore size and surface hydrophilicity of the membrane has not yet been fully explained by researchers. Despite current developments in membrane technology, the practical wastewater applications of PNCMs do not yet have the potential to fully meet expectations. Most of the current investigations on applications of PNCMs are at laboratory scale and unfortunately there are limited studies for industrial production and application. For practical and safe applications, further studies are required to produce economic and industrial-scale membranes.

Conflict of interest

The authors declare no competing interest.

Author details

Muharrem Ince[1*] and Olcay Kaplan Ince[2]

1 Department of Food Processes, Tunceli Vocational School, Munzur University, Tunceli, Turkey

2 Faculty of Fine Arts, Design and Architecture, Department of Gastronomy and Culinary Arts, Munzur University, Tunceli, Turkey

*Address all correspondence to: muharremince@munzur.edu.tr

IntechOpen

© 2022 The Author(s). Licensee IntechOpen. This chapter is distributed under the terms of the Creative Commons Attribution License (http://creativecommons.org/licenses/by/3.0), which permits unrestricted use, distribution, and reproduction in any medium, provided the original work is properly cited. CC BY

References

[1] Ong CS, Goh PS, Lau WJ, Misdan N, Ismail AF. Nanomaterials for biofouling and scaling mitigation of thin film composite membrane: A review. Desalination. 2016;**393**:2-15

[2] Shahmirzadi MAA, Hosseini SS, Tan NR. Enhancing removal and recovery of magnesium from aqueous solutions by using modified zeolite and bentonite and process optimization. Korean Journal of Chemical Engineering. 2016;**33**:3529-3540

[3] Ang WL, Mohammad AW, Hilal N, Leo CP. A review on the applicability of integrated/hybrid membrane processes in water treatment and desalination plants. Desalination. 2015;**363**:2-18

[4] Greenlee LF, Lawler DF, Freeman BD, Marrot B, Moulin P. Reverse osmosis desalination: Water sources, technology, and today's challenges. Water Research. 2009;**43**: 2317-2348

[5] Singh N, Nagpal G, Agrawal S. Water purification by using adsorbents: A review. Environmental Technology and Innovation. 2018;**11**:187-240

[6] Abdali N, Marjani A, Heidary F, Adimi M. Fabrication of PVA coated PES/PVDF nanocomposite membranes embedded with in situ formed magnetite nanoparticles for removal of metal ions from aqueous solutions. New Journal of Chemistry. 2017;**41**:6405-6414

[7] Lin W-C, Li Z, Burns MA. A drinking water sensor for lead and other heavy metals. Analytical Chemistry. 2017;**89**: 8748-8756

[8] Gita S, Hussan A, Choudhury T. Impact of textile dyes waste on aquatic environmentsand its treatment. Environment & Ecology. 2017;**35**: 2349-2353

[9] Kendrick MA. Halogens in Seawater, Marine Sediments and the Altered Oceanic Lithosphere. Switzerland AG: Springer Nature; 2018

[10] Nasir A, Masood F, Yasin T, Hameed A. Progress in polymeric nanocomposite membranes for wastewater treatment: Preparation, properties and applications. Journal of Industrial and Engineering Chemistry. 2019;**79**:29-40

[11] Pandey N, Shukla S, Singh N. Water purification by polymer nanocomposites: An overview. Nano. 2017;**3**:47-66

[12] Inamuddin, Mohd Imran A, Lichtfouse E. Water Pollution and Remediation: Heavy Metals. Switzerland AG: Springer Nature, Springer Science and Business Media LLC; 2021. p. 53

[13] Ali S, Rehman SAU, Shah IA, Farid MU, An AK, Huang H. Efficient removal of zinc from water and wastewater effluents by hydroxylated and carboxylated carbon nanotube membranes: Behaviors and mechanisms of dynamic filtration. Journal of Hazardous Materials. 2019;**365**:64-73

[14] Mahmood Q, Asif M, Shaheen S, Hayat MT, Ali S. Cadmium contamination in water and soil. In: Cadmium Toxicity and Tolerance in Plants from Physiology to Remediation. United States of America: Academic Press; 2019. pp. 141-161

[15] Mezher T, Fath H, Abbas Z, Khaled A. Techno-economic assessment and environmental impacts of desalination technologies. Desalination. 2011;**266**:263-273

[16] Hosseini SS, Bringas E, Tan NR, Ortiz I, Ghahramani M, Shahmirzadi MAA. Recent progress in

development of high performance polymeric membranes and materials for metal plating wastewater treatment: A review. Journal of Water Process Engineering. 2016;**9**:78-110

[17] Li D, Yan Y, Wang H. Recent advances in polymer and polymer composite membranes for reverse and forward osmosis processes. Progress in Polymer Science. 2016;**61**:104-155

[18] Yin J, Deng B. Polymer-matrix nanocomposite membranes for water treatment. Journal of Membrane Science. 2015;**479**:256-275

[19] Liu L, Kentish SE. Pervaporation performance of crosslinked PVA membranes in the vicinity of the glass transition temperature. Journal of Membrane Science. 2018;**553**:63-69

[20] Nataraj D, Sakkara S, Meghwal M, Reddy N. Crosslinked chitosan films with controllable properties for commercial applications. International Journal of Biological Macromolecules. 2018;**120**:1256-1264

[21] Le NL, Nunes SP. Materials and membrane technologies for water and energy sustainability. Sustainable Materials and Technologies. 2016;7:1-28

[22] Wu Y, Liu X, Cui J, Meng M, Dai J, Li C, et al. Bioinspired synthesis of high-performance nanocomposite imprinted membrane by a polydopamine-assisted metal-organic method. Journal of Hazardous Materials. 2017;**323**: 663-673

[23] Unuabonah E, Taubert A. Clay–polymer nanocomposites (CPNs): Adsorbents of the future for water treatment. Applied Clay Science. 2014;**99**:83-92

[24] Taimur S, Yasin T. Influence of the synthesis parameters on the properties of amidoxime grafted sepiolite nanocomposites. Applied Surface Science. 2017;**422**:239-246

[25] Wang C, Yin J, Wang R, Jiao T, Huang H, Zhou J, Zhang L, Peng Q. Facile preparation of self-assembled polydopamine-modified electrospun fibers for highly effective removal of organic dyes. Nanomaterials. 2019;**9**(1):116

[26] Opoku F, Kiarii EM, Govender PP, Mamo MA. Metal Oxide Polymer Nanocomposites in Water Treatments. United Kingdom: IntechOpen Limited; 2017. pp. 173-199

[27] Rana S, Varadwaj GBB, Jonnalagadda S. Ni nanoparticle supported reduced graphene oxide as a highly active and durable heterogeneous material for coupling reactions. Nanoscale Advances. 2019;**1**:1527-1530

[28] Bassyouni M, Abdel-Aziz MH, Zoromba MS, Abdel-Hamid SMS, Drioli E. A review of polymeric nanocomposite membranes for water purification. Journal of Industrial and Engineering Chemistry. 2019;**73**:19-46

[29] Lau WJ, Ismail AF, Misdan N, Kassim MA. A recent progress in thin film composite membrane: A review. Desalination. 2012;**287**:190-199

[30] Yang Y, Gao X, Li Z, Wang Q, Dong S, Wang X, et al. Porous membranes in pressure-assisted forward osmosis: Flux behavior and potential applications. Journal of Industrial and Engineering Chemistry. 2018;**60**:160-168

[31] El-Gendi A, Samhan FA, Ismail N, El-Dein LAN. Journal of Industrial and Engineering Chemistry. 2018;**65**:127-136

[32] Rozelle L, Cadotte J, Corneliussen R, Erickson E, Cobian K, Kopp JRC. Phase inversion membranes. In: Mulder M, editor. Encyclopedia of Separation

Science. NY: Academic Press; 2000. pp. 3331-3346

[33] La Y-H, Sooriyakumaran R, Miller DC, Fujiwara M, Terui Y, Yamanaka K, et al. Novel thin film composite membrane containing ionizable hydrophobes: pH-dependent reverse osmosis behavior and improved chlorine resistance. Journal of Materials Chemistry. 2010;**20**:4615-4620

[34] Wang LK, Chen JP, Hung Y, Shammas NK. Membrane and Desalination Technologies. Switzerland AG: Springer Nature; 2008

[35] Idris A, Man Z, Maulud A, Khan M. Effects of phase separation behavior on morphology and performance of polycarbonate membranes. Membranes. 2017;7:21

[36] Aburabie J, Villalobos LF, Peinemann KV. Composite membrane formation by combination of reaction-induced and nonsolvent-induced phase separation. Macromolecular Materials and Engineering. 2017;**302**:1700131

[37] Amin M. Methods for preparation of nano-composites for outdoor insulation applications. Reviews on Advanced Materials Science. 2013;**34**:173-184

[38] Tanahashi M. Development of fabrication methods of filler/polymer nanocomposites: With focus on simple melt-compounding-based approach without surface modification of nanofillers. Materials. 2010;**3**:1593-1693

[39] Anandhan S, Bandyopadhyay S. Polymer Nanocomposites: From Synthesis to Applications. United Kingdom: IntechOpen; 2011

[40] Müller K, Bugnicourt E, Latorre M, Jorda M, Echegoyen Sanz Y, Lagaron J, et al. Review on the processing and properties of polymer nanocomposites and nanocoatings and their applications in the packaging, automotive and solar energy fields. Nanomaterials. 2017;7:74

[41] Ahmed FE, Lalia BS, Hashaikeh R. A review on electrospinning for membrane fabrication: Challenges and applications. Desalination. 2015;**356**:15-30

[42] Khajavi R, Abbasipour M. Controlling Nanofiber Morphology by the Electrospinning Process. United Kingdom: Elsevier; 2017

[43] Khorshidi B, Tundat T, Pernitsky D, Sadrzadeh M. A parametric study on the synergistic impacts of chemical additives on permeation properties of thin flm composite polyamide membrane. Journal of Membrane Science. 2017;**535**:248-257

[44] Homaeigohar S, Elbahri M. Nanocomposite electrospun nanofiber membranes for environmental remediation. Materials. 2014;7:1017-1045

[45] Hoseinzadeh E, Makhdoumi P, Taha P, Hossini H, Stelling J, Amjad KM. A review on nano-antimicrobials: Metal nanoparticles, methods and mechanisms. Current Drug Metabolism. 2017;**18**:120-128

[46] Besinis A, De Peralta T, Handy RD. The antibacterial effects of silver, titanium dioxide and silica dioxide nanoparticles compared to the dental disinfectant chlorhexidine on Streptococcus mutans using a suite of bioassays. Nanotoxicology. 2014;**8**:1-16

[47] Sadiq IM, Chowdhury B, Chandrasekaran N, Mukherjee A. Antimicrobial sensitivity of *Escherichia coli* to alumina nanoparticles. Nanomedicine: Nanotechnology, Biology, and Medicine. 2009;**5**:282-286

[48] Pramila P, Gopalakrishnan N. Enhancement of antibacterial activity in nanofillers incorporated PSF/PVP membranes. Materials Research Express. 2018;**5**:045306

[49] Mohammad F, Al-Lohedan HA, Al-Haque HN. Chitosan-mediated fabrication of metal nanocomposites for

enhanced biomedical applications. Advanced Materials Letters. 2017;**8**(2): 89-100

[50] Bao C, Guo Y, Song L, Hu Y. Poly(vinyl alcohol) nanocomposites based on graphene and graphite oxide: A comparative investigation of property and mechanism. Journal of Materials Chemistry. 2011;**21**:13942-13950

[51] Lata S, Samadder SR. Removal of arsenic from water using nano adsorbents and challenges: A review. Journal of Environmental Management. 2016;**166**:387-406

[52] Shen C, Wang Y, Xu J, Luo G. Chitosan supported on porous glass beads as a new green adsorbent for heavy metal recovery. Chemical Engineering Journal. 2013;**229**:217-224

[53] Ince M, Kaplan IO. Application of response surface methodological approach to optimize removal of Cr ions from ındustrial wastewater. Atomic Spectroscopy. 2019;**40**(3):91-97

[54] Pala A, Serdar O, Ince M, Onal A. Modeling approach with Box-Behnken design for optimization of Pb bioaccumulation parameters in *Gammarus pulex* (L., 1758). Atomic Spectroscopy. 2019;**40**(3):98-103

[55] Serdar O, Pala A, Ince M, Onal A. Modelling cadmium bioaccumulation in *Gammarus pulex* by using experimental design approach. Chemistry and Ecology. 2019;**35**(10):922-936

[56] Basile A, Cassano A, Rastogi NK. Advances in Membrane Technologies for Water Treatment: Materials, Processes and Applications. Cambridge, United Kingdom: Elsevier Science; 2015

[57] Baker RW. Membrane Technology and Applications. Oxford, UK: Wiley; 2012

[58] Nedzarek A, Drost A, Harasimiuk FB, Tórz A. The influence of pH and BSA on the retention of selected heavy metals in the nanofiltration process using ceramic membrane. Desalination. 2015;**369**:62-67

[59] Shukla AK, Alam J, Ansari MA, Alhoshan M, Alam M, Kaushik A. Selective ion removal and antibacterial activity of silver-doped multi-walled carbon nanotube/polyphenylsulfone nanocomposite membranes. Materials Chemistry and Physics. 2019;**233**: 102-112

[60] Delavar M, Bakeri G, Hosseini M. Fabrication of polycarbonate mixed matrix membranes containing hydrous manganese oxideand alumina nanoparticles for heavy metaldecontamination: Characterization and comparative study. Chemical Engineering Research and Design. 2017;**120**:240-253

[61] Gebru KA, Das C. Removal of chromium (VI) ions from aqueous solutions using amine-impregnated TiO_2 nanoparticles modified cellulose acetate membranes. Chemosphere. 2018;**191**:673e684

[62] Bera A, Trivedi JS, Kumar SB, Chandel AKS, Haldar S, Jewrajka SK. Anti-organic fouling and anti-biofouling poly(piperazineamide) thin film nanocomposite membranes for low pressure removal of heavy metal ions. Journal of Hazardous Materials. 2018;**343**:86-97

[63] Deng S, Liu X, Liao J, Lin H, Liu F. PEI modified multiwalled carbon nanotube as a novel additive in PAN nanofiber membrane for enhanced removal of heavy metal ions. Chemical Engineering Journal. 2019;**375**:122086

[64] Ghaemi N, Zereshki S, Heidari S. Removal of lead ions from water using PES-based nanocomposite membrane incorporated with polyaniline modified GO nanoparticles: Performance optimization by central composite design.

Process Safety and Environmental Protection. 2017;**111**:475-490

[65] Gohari RJ, Lau WJ, Matsuura T, Halakoo E, Ismail AF. Adsorptive removal of Pb(II) from aqueous solution by novel PES/HMO ultrafiltration mixed matrix membrane. Separation and Purification Technology. 2013;**120**: 59-68

[66] Hofman M, Pietrzak R. Copper ions removal from liquid phase by polyethersulfone (PES) membranes functionalized by introduction of carbonaceous materials. Chemical Engineering Journal. 2013;**215-216**: 216-221

[67] Ibrahim GPS, Isloor AM, Asiri AM, Ismail AF, Kumar R, Ahamed MI. Performance intensification of the polysulfone ultrafiltration membrane by blending with copolymer encompassing novel derivative of poly(styrene-comaleic anhydride) for heavy metal removal from wastewater. Chemical Engineering Journal. 2018;**353**:425-435

[68] Karim Z, Mathew AP, Kokol V, Wei J, Grahn M. High-flux affinity membranes based on cellulose nanocomposites for removal of heavy metal ions from industrial effluents. RSC Advances. 2016;**6**:20644-20653

[69] Liu C, Lei X, Wang L, Jia J, Liang X, Zhao X, et al. Investigation on the removal performances of heavy metal ions with the layer-by-layer assembled forward osmosis membranes. Chemical Engineering Journal. 2017;**327**:60-70

[70] Liu X, Wu J, Wang J. Removal of Cs(I) from simulated radioactive wastewater by three forward osmosis membranes. Chemical Engineering Journal. 2018;**344**:353-362

[71] Lv Y, Du Y, Chen Z-X, Qiu W-Z, Xu Z-K. Nanocomposite membranes of polydopamine/electropositive nanoparticles/polyethyleneimine for nanofiltration. Journal of Membrane Science. 2018;**545**:99-106

[72] Masheane ML, Nthunya LN, Malinga SP, Nxumalo EN, Mamba BB, Mhlanga SD. Synthesis of Fe-Ag/f-MWCNT/PES nanostructured-hybrid membranes for removal of Cr(VI) from water. Separation and Purification Technology. 2017;**184**:79-87

[73] Saeedi-Jurkuyeh A, Jafari AJ, Kalantary RR, Esrafili A. A novel synthetic thin-film nanocomposite forward osmosis membrane modified by graphene oxide and polyethylene glycol for heavy metals removal from aqueous solutions. Reactive and Functional Polymers. 2020;**146**:104397

[74] Seyedpour SF, Rahimpour A, Mohsenian H, Taherzadeh MJ. Low fouling ultrathin nanocomposite membranes for efficient removal of manganese. Journal of Membrane Science. 2018;**549**:205-216

[75] Shah P, Murthy CN. Studies on the porosity control of MWCNT/polysulfone composite membrane and its effect on metal removal. Journal of Membrane Science. 2013;**437**:90-98

[76] Ye C-C, An Q-F, Wu J-K, Zhao F-Y, Zheng P-Y, Wang N-X. Nanofiltration membranes consisting of quaternized polyelectrolyte complex nanoparticles for heavy metal removal. Chemical Engineering Journal. 2019;**359**:994-1005

[77] You S, Lu J, Tang CY, Wang X. Rejection of heavy metals in acidic wastewater by a novel thin-film inorganic forward osmosis membrane. Chemical Engineering Journal. 2017;**320**:532-538

[78] Youssef AM, El-Naggar ME, Malhat FM, El Sharkawi HM. Efficient removal of pesticides and heavy metals from wastewater and the antimicrobial activity of f-MWCNTs/PVA nanocomposite film. Journal of Cleaner Production. 2019;**206**:315-325

Section 2

Recent Trends in Membranes and Membrane Processes

Chapter 2

Shale Oil and Gas Produced Water Treatment: Opportunities and Barriers for Forward Osmosis

Layla Ogletree, Hongbo Du and Raghava Kommalapati

Abstract

The treatment of shale oil & gas produced water is a complicated process since it contains various organic compounds and inorganic impurities. Traditional membrane processes such as reverse osmosis and nanofiltration are challenged when produced water has high salinity. Forward osmosis (FO) and membrane distillation as two emerging membrane technologies are promising for produced water treatment. This chapter will focus on reviewing FO membranes, draw solute, and hybrid processes with other membrane filtration applied to produced water treatment. The barriers to the FO processes caused by membrane fouling and reverse draw solute flux are discussed fully by comparing some FO fabrication technologies, membrane performances, and draw solute selections. The future of the FO processes for produced water treatment is by summarizing life cycle assessment and economic analyses for produced water treatment in the last decade.

Keywords: forward osmosis, produced water, desalinization, treatment technology, draw solution

1. Introduction

Shale oil & gas produced water (PW) is the wastewater generated from the oil and gas industry and is a significant portion of the industrial wastewater. During the exploration and production of oil and gas, water is injected with some chemicals to cause a hydraulic fracture, and this water is trapped underground. Then this water is brought to the ground surface and is referred to as PW. Because of the hydrofracturing conditions, the mixed water's characteristics and properties have been altered physically and chemically [1]. These characteristics vary depending on location, time, the drilling, production, treatment processes, and the geological formation in contact with the injected water. The produced water characteristics are numerous, but the most prominent are salt, oil, naturally occurring radioactive materials (NORM), organic and inorganic chemicals, metals, and various suspended and total dissolved solids [1–3]. Because of these properties, PW treatment is a very strenuous and costly process. The PW treatments can be broken down into three main stages: primary, secondary, and tertiary.

Produced water is characterized as high salinity, which implies a high concentration of total dissolved solids (TDS) (sodium, chloride, calcium, magnesium, sulfate, and other dissolvable ions or solids) [4]. Besides free, dispersed, or emulsified oil droplets present in PW, other characteristics are calcium & magnesium causing hardness, phenols, benzene, toluene, ethylbenzene xylenes (BTEX), acids, non-volatile, volatile, and semi-volatile organics, and other chemical additives. With a wide variety of contaminants within PW, various treatment methods are required for cleaning PW instead of the deep-well injection by one or several thousand feet deep underground.

Among many processes used to cleanse PW, desalination, in which salt and other minerals are removed from the water to create freshwater, has become a forerunner. The primary method used for desalination is reverse osmosis (RO). Osmotically driven membrane processes (ODMPs), specifically forward osmosis (FO), are the inverse of pressure-driven membrane processes (PDMP), e.g., ultrafiltration, nanofiltration, reverse osmosis (RO). Pressure-driven membrane processes (PDMPs) are when high pressure is used to drive water flow through a semipermeable membrane. Inversely, ODMPs are caused by an osmotic pressure difference that pushes water flow through a semipermeable membrane from the dilute feed water (FW) to a concentrated draw solution (DS). Comparatively, ODMPs have several benefits and advantages that PDMP does not, such as low energy consumption, ease of equipment maintenance, low capital investment, high salt rejection, and high-water flux. Because of the cost and energy required in the RO systems, FO systems are being considered. FO also offers benefits such as higher productivity, rejection of various contaminants, and lower fouling tendencies.

2. Produced water

The oil and gas industry produces a lot of waste, 80% of which is a liquid waste [1]. In 2012, it was reported that the oil and gas industry in the US produced approximately 3.4 trillion liters, or 0.88 trillion gallons of PW yearly, which was 2.46 billion gallons per day [3, 4]. This number is projected to continue to increase by 25% from 2006 to 2030 [1]. The oil and gas industry not only produces wastewater but uses water in several of its processes. The water produced may be discharged into the surrounding environment but must meet that region's discharge requirements. It may also be used for consumption though this would require extensive treatment.

2.1 Characteristics of produced water

There are various contaminants in PW that depend on the geological location, the geologic formation where the water is generated, and the hydrocarbon product produced. These contaminants can be organized into sections below [1–3]:

- Dispersed oil
- Dissolved oil
- Dissolved minerals & metals
- Dissolved gases
- Production chemicals
- Produced solids & bacteria

Dispersed oil and dissolved organic compounds categories are frequently grouped due to their similarities. Dispersed oil refers to the dispersed oil droplets suspended in PW and can become toxic at the point of discharge [2]. These droplets are usually four to six microns but can fluctuate from two to 30 microns and are originated from the aqueous phase. Dispersed oil contributes considerably to the biological oxygen demand (BOD) and would cause an issue with the use of PW for consumption or discharging into nearby ecosystems. Dissolved oil refers to the organic compounds found in PW, and it includes phenols, BTEX, volatile hydrocarbons, polyaromatic hydrocarbons (PAHs), phenols, carboxylic acid, and other organic acids [1, 2]. These components are found in higher concentrations and are created from the platforms that produce oil rather than the plan platforms that produce gas [2]. **Table 1** illustrates the typical composition of PW samples from natural gas production.

Dissolved minerals are many inorganic compounds that naturally occur in PW. These compounds are found at high concentrations within PW and can be classified as cations and anions. Dissolved minerals can also precipitate to form solids, affecting TDS, which hinder the treatment process [2]. These inorganic compounds include heavy metals and NORMs and are typically less toxic when compared to their organic counterparts. The concentrations of various mineral ions present in shale gas PW can be seen in **Table 2**.

The main metal and minerals contributing to the high concentration of dissolved minerals are sodium, chloride, sulfate, bicarbonate, aluminum, strontium, potassium, calcium, iron, barium, and magnesium [9]. These minerals are the main contributor to the inorganic chemistry of PW, with sodium and chloride affecting its salinity and chloride, sodium, potassium, strontium, calcium, barium, magnesium, iron, sulfate, carbonate, and bicarbonate affecting its conductivity and scale-forming potential [2]. Radium-226 and radium-228 are the most abundant NORMs in PW, with barium sulfate as its frequently co-precipitated scale. Dissolved gases refer to the gases that occur naturally due to bacterial activity or chemical reactions within PW. These gases commonly are carbon dioxide, oxygen, and hydrogen sulfide [2].

Production chemicals refer to the chemicals added during oil and gas production to further aid in operation and production. These chemicals can be divided into two sections treatment chemicals and production treating chemicals. These chemicals

Parameter	Unit	Minimum value	Maximum value
pH		3.1	7.0
Conductivity	mg L^{-1}	4,200	180,000
TOC	mg L^{-1}	67	38,000
TDS	mg L^{-1}	2,600	310,000
TSS	mg L^{-1}	8	5,484
BOD_5	mg L^{-1}	75	2,870
COD	mg L^{-1}	2,600	120,000
Oil/grease	mg L^{-1}	2.3	60
Surfactants	mg L^{-1}	Not detected	285
Alkalinity	mg L^{-1}	Not detected	285

Table 1.
Composition of produced water [1, 2, 4–8].

Metal/ Mineral	Minimum value (mg L^{-1})	Maximum value (mg L^{-1})
Aluminum	Not detected	83
Arsenic	0.004	151
Barium	Not detected	1740
Boron	Not detected	56
Bromide	150	1149
Cadmium	Not detected	1.21
Calcium	Not detected	51,300
Chloride	1400	190,000
Chromium	Not detected	0.03
Copper	Not detected	5
Iron	Not detected	1100
Lead	<0.2	10.2
Lithium	18.6	235
Magnesium	0.9	4300
Manganese	0.045	63
Nickel	Not detected	9.2
Potassium	149	3870
Sodium	520	120,000
Strontium	Not detected	6200
Sulfate	Not detected	47
Silver	0.047	7
Tin	Not detected	1.1
Zinc	Not detected	5

Table 2.
Minerals in produced water [2, 6–8].

can be either pure compounds or compounds that contain active ingredients dissolved in a solvent or a co-solvent [1]. Treatment chemicals are used in production treating, gas processing, and stimulation. Production treating chemicals are used in scale and corrosion inhibitors, biocides, emulsion breakers, antifoam, and water treatment chemicals. These chemicals can be seen in **Table 3.**

Solids vary from oilfield to platform and refer to the myriad of constituents produced from the formation and other operations. These solids include scale products, sand and silt, carbonates, clay, wax, corrosion products, and other suspended solids. They may consist of bacteria, which can create sulfides when the bacterial reduction of sulfate occurs [4]. However, the quantity of bacteria is usually relatively small due to various toxic chemicals within PW [1, 4]. These microorganisms found in PW are typically aerobic and Gram-positive. These solids can hinder oil production due to the oily sludge and emulsions that can clog flow lines.

2.2 Treatment of produced water

Currently, there is a myriad of treatments used for PW. The goal of these treatments are; (i) de-oiling (removing dispersed and dissolved oil), (ii) soluble

Chemical	Concentration: Oilfield		Concentration: Gas Field	
	Typical ($mg\,L^{-1}$)	Range ($mg\,L^{-1}$)	Typical ($mg\,L^{-1}$)	Range ($mg\,L^{-1}$)
Corrosion Inhibitor	4	2–10	4	2–10
Scale Inhibitor	10	4–30	—	—
Demulsifier	1	1–2	—	—
Polyelectrolyte	2	0–10	—	—
Methanol	—	—	2000	1000–15000
Glycol (DEG)	—	—	1000	500–2000

Table 3.
PW treatment chemicals [1, 9, 10].

organics removal, (iii) disinfection, (iv) removal of suspended solids, (v) removal of dissolved gas (removing of light hydrocarbon gases, carbon dioxide, and hydrogen sulfide), (vi) desalination, (vii) softening, and (viii) removing NORM. The traditional treatment processes of PW can be separated by the categories:

- Physical treatment
- Chemical treatment
- Biological treatment

The main physical treatment methods for PW are physical adsorption, sand filters, cyclones, evaporation, electrodialysis (ED), dissolved air precipitation (DAP), and freeze–thaw/evaporation (FTE). The main chemical treatment methods for PW are chemical precipitation, chemical oxidation, electrochemical process, photocatalytic treatment, Fenton process, treatment with ozone, room temperature ionic liquids, and demulsifiers. The biological treatment methods for PW are the use of aerobic and anaerobic microorganisms.

Treatments outside of these three main categories are thermal and membrane filtration technologies. Thermal technology is unique to regions where the cost of energy is low-priced. Before membrane technologies reached the forefront, thermal treatment technologies were the main methods used in desalination. The main thermal technology methods for PW are multistage flash (MSF) distillation, vapor compression distillation (VCD), and multi-effect distillation (MED) [11]. Another thermal technology method is a hybrid of multi-effect distillation and vapor compression distillation (MED-VCD hybrid).

Membrane treatment technology has become a promising method for PW desalination [12]. The main membrane filtration methods for PW are microfiltration (MF), ultrafiltration (UF), nanofiltration (NF), and reverse osmosis (RO) membranes. These can either be pressure or vacuum-driven systems with either a crossflow filtration or dead-end filtration system [1]. These membranes are microporous films made of synthetic organic or inorganic materials with various pore sizes [1, 2, 4]. The advantages of membrane filtration technology are sludge reduction, high quality of permeate, less space required, easiness of operation, minimal impact on permeate quality with variation in feed water quality, little to no chemicals needed, the possibility for recycling waste streams, the potential for having an automated plant, moderate capital costs, ability to be combined easily with other separation processes, low energy consumption, and continuous separation [2].

Among different types of membranes, MF membranes have the largest pores size from 0.1 to 5 μm and remove suspended solids and turbidity. The pore size of UF membranes range from 0.01 and 0.1 μm and these membranes remove color, odor, viruses, and colloidal organic matter. It is one of the most effective treatment methods for the removal of oil from PW. Both MF and UF membranes are made of polymeric and inorganic materials. Comparatively, NF and RO membranes are made of polymeric materials when compared to the more porous UF and MF membranes. The pore size of NF membranes range from 0.5 to 1 nm and these membranes remove multivalent ions and specifically charged or polar molecules. Additionally, RO membranes remove low molecular weight components.

3. Forward osmosis membranes

Forward osmosis is a membrane filtration process that has drawn interest in the PW treatment. The FO method is mainly used for desalination due to its ability to remove salt, minerals, and other compounds from water efficiently [13, 14]. This is because the process of FO is based on a phenomenon that occurs naturally. This phenomenon is when two solutions of varying concentrations are separated by a semipermeable membrane, and the solution with the lower concentration will pass through the membrane to the solution with the higher concentration [15]. The FO process relies on the osmotic pressure difference that causes the aforementioned phenomenon to complete the desalination task without the need of an external pressure source, unlike RO. The FO process may be used standalone for desalination or in an enhanced RO pretreatment process [13]. In FO, there are two solutions, a draw solution (DS), with high salinity and osmotic pressure, and a feed solution (FS), with low salinity and osmotic pressure, are separated by a semipermeable membrane and use the natural osmotic pressure difference to move the water from FS to DS until equilibrium is reached [16, 17]. Feed solution will increase in concentration as DS dilutes.

In general, FO membranes are asymmetric and consist of two layers, an active layer and a support layer, as shown in **Figure 1**. The most main commercially available of the various membranes used in FO are cellulose acetate (CA), cellulose triacetate (CTA), and thin-film composite (TFC) membranes [14, 19].

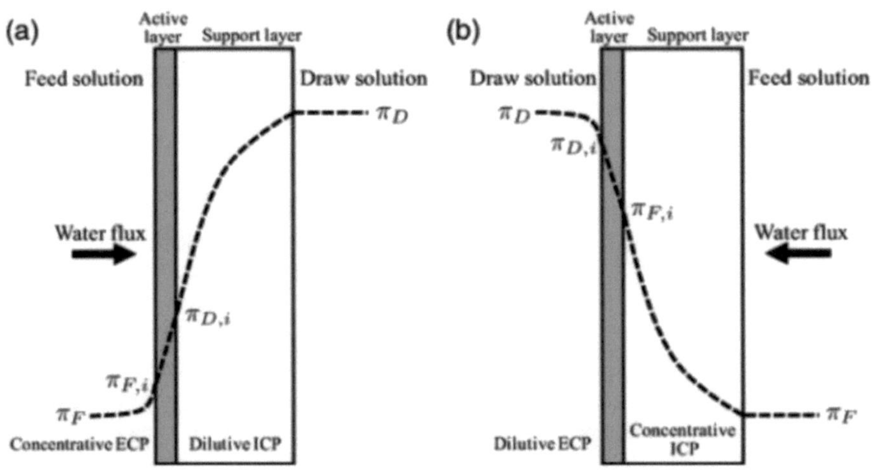

Figure 1.
Principle of forward osmosis (FO). (a) AL-FS mode, and (b) AL-DS mode [18].

Its primary purpose of the support layer is to provide mechanical support to the active layer. One imperative aspect that the support layer directly affects is internal concertation polarization (ICP). The IPC phenomenon, as to be discussed further later, is when the osmotic driving force is impaired, in this case, due to dissonance between both the active and support layer [20]. When this occurs, it frequently causes the water flux to decrease. This disconnect between the two layers is when there is a resistance in the transfer of masses from one layer to the other.

Initially, FO membranes were produced using CA, CTA, polysulfone (PS), or polyethersulfone (PES), and polybenzimidazole (PBI). And in later years, TFC membranes were used on polysulfone, sulfonated, cellulose acetate propionate (CAP), and non-woven electrospun nanofibrous substrates [16, 18, 21, 22]. The phenomenon that is osmosis was first observed in 1748 by Nollet, but no process was made towards the synthesis of these membranes until years later [23, 24]. CA membranes were first hypothesized in 1957, and a case study sheds light on the possibilities this semipermeable membrane contains due to its behavior in saline water [16]. The first synthetic membrane was created in 1963 by Loeb and Sourirajan [16, 23]. This membrane was an asymmetric RO membrane, fabricated from CA, acetone, and aqueous magnesium perchlorate, with high flux and high salt rejection [16, 23]. Following this, in 1965, Batchelder utilized natural cellulose as the foundation for a semipermeable FO membrane [24].

Further employing dissolved volatile solutes, such as sulfur dioxide in seawater or fresh water, as the DS and seawater as the FS [16]. Several more studies were conducted, and in 1986 the Hydration Technology Innovation (HTI) was able to fabricate a CTA FO membrane for commercialization [25]. This membrane was used for various things, most notably as a water filtration system for global military forces and humanitarian disaster relief organizations especially [16, 23]. A new FO desalination process came to fruition in 2005, where ammonium bicarbonate was used as DS to create a high osmotic gradient across the FO membrane, and a new step was added [26]. Upon a moderate increase of temperature, ammonium bicarbonate decomposes into ammonia and carbon dioxide gases that can be isolated and reused as DS, leaving fresh PW [16]. The step added to this FO process, named indirect osmosis, is the DS regeneration, where after the DS is diluted and the concentrated DS is combined with the diluted DS [24, 25].

Over the years, more FO membranes were fabricated, tested, and made available for commercial use. CA and CTA membranes are some of the original materials used for FO applications. These membranes are still created and commercially sold today. They are not sufficient for wastewater treatment due to their operational efficiency parameters. These commercialized CA and CTA membranes provide low permeability, low salt rejection, poor resistance to biological species, and limited chemical stability [27]. PS, or PES, is another material use to fabricate FO membranes. The operation parameters for PS FO membranes are hindered because of the lack of durability and filtering capabilities. There is currently research on how to remedy these concerns using nanomaterials, such as carbon nanotubes and graphene oxide, to strengthen the membrane and increase its durability. Another important material for FO membranes is PBI. The PBI FO membranes are highly hydrophobic, and this affects their ability to efficiently be applied to water purification. Membranes that are more efficient and are more favored are polyamide fabricated. These membranes are fabricated using TFC and interfacial polymerization (IP) methods. These membranes are preferred because of their ability and because their operational efficiency parameters are much broader. They can be used under much more strenuous conditions and constraints.

3.1 Advantages of FO membranes

The utilization of FO for desalination offers a myriad of benefits when compared with pressure-driven membrane processes. One of the main benefits is the ability of FO membranes for low energy consumption due to not requiring applied hydraulic pressure. Thus, in turn, reducing the cost, especially if the DS is economically and technically viable [15]. However, this factor is contentious due to the energy required to regenerate the DS. The amount of energy utilized is comparable to the energy used in RO processes. However, this cost can be reduced by selecting a low-cost energy source for the FO process. Additionally, FO processes can be modified and integrated with other processes to reduce energy costs. Hybrid FO processes will be discussed further in Section 4.3.

Another benefit of FO that is tied to not requiring hydraulic pressure is a reduction of fouling. Recent studies have demonstrated that a lack of applied hydraulic pressure and optimization of the system's hydrodynamics can lead to less fouling and the higher reversibility of fouling [15, 28–31]. Due to the low flux that can be obtained in FO, there is more opportunity for low fouling [27, 31, 32]. Because of this, a more extensive range of pollutants can be removed from FS. The reduction of fouling can also be attributed to FO fouling layers being less dense than the fouling layers of its pressure-driven counterparts, also assisting in its reversibility. The FO processes are capable of treating FS with high fouling propensity due to their lower fouling tendency without lengthy pretreatment. The lower the fouling tendency of the membrane, the longer the lifetime of the membrane [33]. Membranes with lower fouling tendencies require cleaning and maintenance less frequently. Furthermore, the fouling on FO membranes can be physically cleaned instead of cleaning with chemicals. An additional benefit of FO is a phenomenon called retarded forward diffusion of solutes [34]. In this phenomenon, the reverse salt flux in FO processes impedes the pore diffusion of feed solutes, leading to the high rejection of feed solutes [28, 34]. As mentioned earlier, when a high osmotic pressure gradient is obtained across the membrane, sufficient water flux and recovery can be achieved [15, 35]. The increase in water recovery would cause a decrease in rejected FS, leading to a reduction in waste disposal. This would be especially beneficial for desalination plants, in particular plants offshore [15].

3.2 Challenges faced by FO membranes

Though there are many advantages to FO membranes, there are still several challenges that limit their applications. These challenges include:

- Concentration polarization
- Membrane fouling
- Reverse solute diffusion

These challenges are caused by membrane orientation and design, DS and FS concentrations, and operational conditions. Membrane orientation refers to the order in which the displacement of the FO membrane can be alternated. The AL-FS mode is where the active layer faces FS, and the AL-DS mode is where the active layer faces DS. Concentration polarization (CP) occurs in all forms of membrane separation processes, as shown in **Figure 2**. In FO membranes, CP occurs because of a concentration gradient between FS and DS through an asymmetric FO membrane [15, 19, 37]. Furthermore, CP can be broken into internal concentration polarization

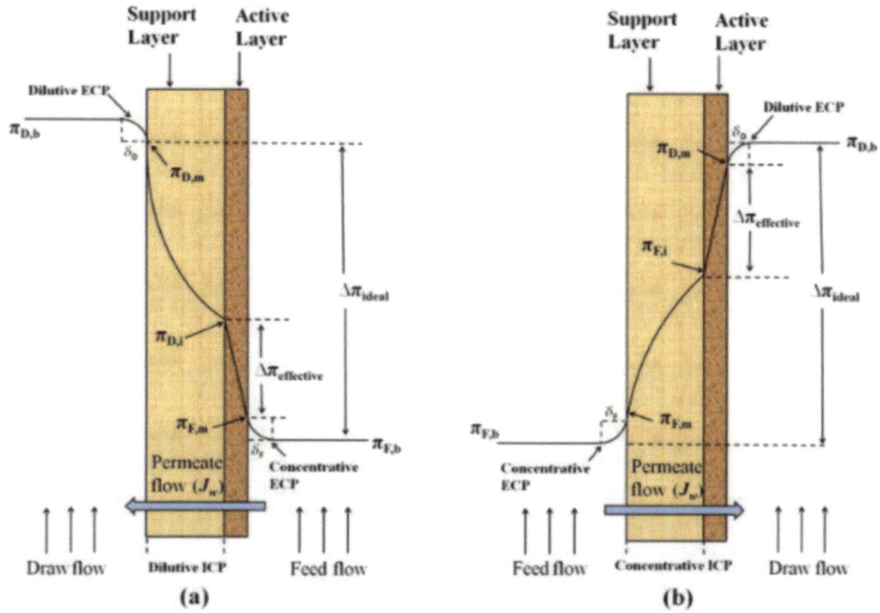

Figure 2.
Internal concentration polarization (ICP) and external concentration polarization (ECP) occurring in forward osmosis. (a) ICP and ECP in Al-FS mode, and (b) ICP and ECP in Al-DS mode [36].

(ICP) and external concentration polarization (ECP). ICP occurs within the FO membrane's support layer, and ECP occurs around the surface of the FO membrane's active layer.

The ECP phenomenon occurs around the surface of the FO membrane's active layer and is caused by a difference in the membrane surface's concentration and the concentration of the bulk solution [36]. Moreover, ECP can be defined further by two types, concentrative and dilutive. Concentrative ECP occurs in AL-FS mode and is when the FS becomes concentrated at the active layer surface as the water travel through the membrane. Dilutive EPC occurs in AL-DS mode and is when the DS is diluted on the active layer surface of the water travels through the membrane. Additionally, EPC can significantly affect the osmotic gradient, playing a pivotal role in its decrease and thus hindering water flux across the membrane [15]. Though, the amount of EPC occurring depends significantly on the DS chosen. Moreover, EPC does not have as great of an impact on the membrane's effectiveness as IPC does.

The ICP phenomenon occurs within the FO membrane's support layer and is when solutes are unable to penetrate the dense active layer, which causes the CP within the porous support layer [15, 36]. ICP can also be defined further by two types, concentrative and dilutive. Concentrative ICP occurs in AL-DS mode and is when the FS concentrates within the support layer as the water travel through the membrane. Dilutive IPC occurs in AL-FS mode and is when the DS is diluted within the support layer as the water travel through the membrane. When concentrative ICP occurs, dilutive ECP. The inverse occurs with dilutive ICP. When dilutive ICP occurs, concentrative ECP occurs.

4. Optimization of forward osmosis membranes for produced water

An ideal FO membrane consists of an active layer that high water permissibility and a support layer that minimizes CP, both internal and external, and thus enhances mass transfer [16]. The FO membrane must also have mechanical stability, chemical resistance and less susceptible to membrane fouling. The FO membranes with these qualities have advantages over other membrane processes, such as UF, MF, NF, and RO [12]. These advantages are the capability of operating at low hydraulic pressure and an ambient temperature whilst rejecting almost all solutes and suspended solids. Because of those operating abilities, energy consumption is reduced. The cost of construction and system procedure is due to FO membranes not requiring high hydraulic pressures to overcome high osmotic pressure [38]. These capabilities translate well into harsh conditions, where there is limited access to electricity, currency, and materials. The effectiveness of the FO membrane can be described by the values of the intrinsic parameters A, B, and S of the membrane [39]. These parameters vary depending on the structure of the membrane and the types of solutes utilized.

The membrane structural parameter, S, is key when determining the performance of FO membrane and performance. The ICP within the support layer of the membrane can be determined using the S parameter as it is a constant which is dependent on the concentrations of the DS and FS. S can be denoted as:

$$S = \left(\frac{A\pi_{ds} + B}{A\pi_{fs} + B + J_w} \right) \frac{D}{J_w} \quad (1)$$

Where A is the water permeability (L m-^{2} h^{-1} bar^{-1},), B is the salt permeability (L m-^{2} h^{-1}), D is the salt diffusion coefficient (m^2 h^{-1}), J_w is the water permeation flux (L m-^{2} h^{-1} or LMH), π_{ds} is the bulk osmotic pressure of the DS (bar), and π_{fs} is the bulk osmotic pressure of the FS (bar). The equation for water permeability, A, can be noted as:

$$A = \frac{J_w}{\Delta P} \quad (2)$$

Where ΔP is the transmembrane pressure. B, the salt permeability can be denoted as [15, 40]:

$$B = \left(\frac{1}{R} - 1 \right) J_w \quad (3)$$

Where R is the salt rejection, which can be denoted by:

$$R = \frac{C_f - C_p}{C_f} \cdot 100\% \quad (4)$$

Where C_f is the salt concentration of the FS (M) and C_p is the concentration of the DS (M). J_w, the water permeation flux, is denoted as:

$$J_w = \frac{Q}{A_m \cdot \Delta t} \quad (5)$$

Where A_m is the effective membrane area (m^2), and Q is the volume of the water permeate (L) that is collected over an elapsed time of Δt (h). In AL-DS mode, also known as PRO mode, J_w can be denoted as [15, 41]:

$$J_w = \frac{\frac{1}{K}\ln\left(A\pi_{ds} + B - J_w\right)}{A\pi_{fs} + B} \tag{6}$$

Where K is the solute resistivity (d m^{-1}), and A is the water permeability. In AL-FS mode, also known as FO mode, J_w can be denoted as [15, 41]:

$$J_w = \frac{\frac{1}{K}\ln\left(A\pi_{ds} + B\right)}{A\pi_{fs} + B + J_w} \tag{7}$$

The solute resistivity (m), K, can be denoted as [15, 41]:

$$K = \frac{t_s\tau}{D\varepsilon} = \frac{S}{D} \tag{8}$$

Where t_s is the thickness of the support layer, ε is the porosity of the membrane, and τ is the tortuosity of the membrane. The reverse salt flux (g m-^{2} h^{-1} or gMH), J_s, can be denoted as [24, 40]:

$$J_s = \frac{\Delta C \cdot \Delta V}{\Delta t \cdot A_m} = \frac{C_t V_t - C_0 V_0}{\Delta t \cdot A_m} = B \cdot \Delta C \tag{9}$$

Where ΔC is the difference in the salt concentration of the FS across the membrane (M) and ΔV is the change of the FS volume.

The parameters can be measured and calculated to exemplify the proficiency of a membrane. There are used widely in membrane trials testing how a variable of the FO membrane may affect the membrane's efficiency.

The membrane's structure begins within its layers and what these layers consist of. As mentioned earlier, FO membranes are mainly porous asymmetric membranes which consist of a dense active layer, with a thickness from 0.1 to 1 μm, that is supported by a highly porous support layer, with a thickness from 100 to 200 μm [19, 42]. The thickness of and the pore size, from 0.4 to 1 nm, the active layer determines the chemical structure of the membrane.

There are many ways to directly measure the intrinsic structural parameter of the membrane. These can be broken into two groups, imaging and analytical characterization techniques. The main imaging characterization techniques use to directly measure the intrinsic structural parameter of the membrane are x-ray tomography, confocal laser scanning microscopy, and scanning electron microscopy. The main analytical characterization techniques used to directly measure the intrinsic structural parameter of the membrane are intrusion and extrusion porosimetry, gravimetric analysis of porosity, and electrochemical impedance spectroscopy. Once these properties of the membrane have been measured or calculated, they can be used to predict other parameters of occurrences. S of the support layer that cannot be lower than the thickness of the support layer can be used to predict ICP. A high value for S indicates a high amount of ICP and thus a low amount of water flux. Therefore, an ideal membrane would have a low S value. Membrane permeance can also be used to indicate water flux. Low permeance, less than 0.5 L m^{-2} h^{-1} bar^{-1}, results in little to no high flux meaning a less ideal membrane. The ideal water permeability of an FO membrane is at least 1 L m^{-2} h^{-1} bar^{-1}. Thus, an ideal membrane would need a medium, 0.5 t0 2.0 L m^{-2} h^{-1} bar^{-1}, or higher than 2.0 L m^{-2} h^{-1} bar^{-1}, value of A [28].

The active layer has several core characteristics, molecular weight, roughness of its surface, and thickness, and determines the separation properties of the membrane. As mentioned earlier, FO membranes require a highly rejecting active layer to preserve a high driving force from its osmotic pressure difference. There are three main methods to fabricate the active layer:

- Interfacial polymerization (IP)
- Layer-by-layer assembly (LbL)
- Conventional phase inversion
- Surface blending & grafting

The IP method is used to fabricate a thin polyamide layer as the active layer. The TFC membranes fabricated via IP have shown high salt rejection and high water permeability capabilities [43]. The IP methodology creates a thin film by dissolving two highly reactive monomers into two immiscible liquids, which in turn causes a reaction [43, 44]. The monomers used in IP are usually nucleophile reactants (typically multi-functional amines) and electrophile reactants (typically multi-functional acid chlorides) [43, 45]. The most popular monomers used are m-phenylenediamine (MPD) and trimesoyl chloride (TMC) [45]. The characteristics of TFC active layers via IP are determined by choice of the monomer, reactivity, diffusivity, concentration, and solubility. The characteristics of the support layer, pore size, porosity, roughness, and hydrophilicity play an important role when considering the characteristics of TFC layer [45]. The ideal support layer surface for IP is hydrophilic and has a pore size from 1 to 100 nm. Additionally, TFC membranes can be improved through the incorporation of carbon nanotubes or aquaporins. The polyamide layer forming through IP is also influenced heavily by the IP's experimental conditions. IP requires high experimental sensitivity and thus is complex to reproduce.

The LbL assembly method is where a thin selective film is formed by placing polyelectrolytes (PE), an oppositely charged layer, on top of a charged layer. The LbL method allows for more control precision with layers, ultra-thin defect-free layers, and various types of PE to be incorporated [28]. Until like IP, LbL assembly is straightforward and does not require complex instruments. LbL membranes can also increase water permeability, thermal stability, and high solvent resistance through the creation of an ultra-thin defect-free layer [46]. LbL membranes have limited options for DSs due to their low rejection of small draw solutes, specifically NaCl [28].

Phase inversion is a method commonly used to fabricate asymmetric membranes. This method is used in the fabrication of most commercially sold membranes. This process begins with liquid–liquid demixing to create a polymer solution. The solution is then altered thermodynamically to create an active layer that is in turn cross-linked to the support layer. The demixing portion of the phase inversion process can be done through several methods: immersion precipitation, controlled evaporation, thermal precipitation, and precipitation from the vapor phase [47].

Surface blending is a method where materials are added to alter and improve the physical and chemical properties of the membrane. Nanoparticles and nanofibers have been incorporated to enhance and strengthen the membrane [48]. Surface grafting is where inorganic nanoparticles or organic functional groups are grafted onto the surface of the support layer. The several methods utilized for surface grafting: plasma discharge, UV irradiation, and ozone [49]. Surface grafting allows for more integration into the polymer's structure and nanocomposite membrane stability [28]. Another method used for asymmetric membranes is phase separation. Phase separation is used for hollow fiber (HF) membrane fabrication. Hollow fiber membrane fabrication is where the active and support layers are fabricated, via spinning, concurrently.

The support layer is usually fabricated using phase inversion, typically immersion precipitation, as discussed earlier. The support layer determines the water flux due to ICP. Because FO membranes do not require external pressure, they require less mechanical strength. Though less mechanical strength is required, the mechanical strength of the membrane is still essential, just less dense. A central part in the fabrication of the support layer is the polymer used in the phase inversion, the composition of the polymer solution, and the coagulation bath [28].

There are three FO membranes that are most commonly used: TFC-FO, asymmetric cellulose triacetate (CTA), and modified thin-film composite (TFC-RO) membranes [26, 50–52]. TFC-FO membranes are currently at the forefront of FO membrane research. The common methods in these studies are IP for the fabrication of the active layer, phase inversion for the support layer (electrospinning, vacuum filtration, immersion precipitation), membrane coating, and the integration of nanoparticles and nanofibers into the membrane [53–59]. Materials commonly used in the processes are N, N-dimethylformamide (DMF) within the support layer, polyamide for the active layer, and NaCl as DS. These vary with various other materials included. There is a myriad of nanoparticles infused into the support layers of the membranes. In one recent study, polyacrylonitrile (PAN) was infused into the support layer of a TFC-FO membrane with PA via IP for the active layer and NaCl as DS [58]. This membrane was able to achieve a lower reverse salt flux, mitigated ICP, and improve its water flux. As of recent, GO is another nanomaterial imbued in TFC-FO membranes. The hydrophilic nature of GO aids in the exchange between FS and DS. A comparison of all GO FO membranes was concluded by Wu et al., where all the data GO membranes were collected and analyzed. All of the membranes in this study showed the capability to improve water permeability, salt rejection efficiency, support mechanical strength, and provided remarkable antifouling properties [60]. Additionally, HF FO membranes are also considered due to their sizable surface area compared to their volume [61]. These membranes have shown reduce ICP and have high porosity and a strong pore structure [62]. The use of nanomaterials within the support layer of various types of FO membranes has with been proven to strengthen the membrane's structure, increase water flux, and reduce ICP [63]. Other materials are being incorporated into FO membranes, such as aquaporin (AQP). These membranes are biometric membranes and standout for their high water permeability [64]. Recent studies on LbL FO membranes have also shown high water permeation flux of 100 LMH, but use 3 M $MgCl_2$ as DS, due to rejection small draw solutes.

4.1 Draw solutions for FO

The DS chosen for the FO process is integral to the overall performance of the membrane. FO membranes are driven by osmotic pressure, so if the incorrect DS is chosen, there may not be enough osmotic pressure. An ideal DS would be stable, inexpensive, and accessible, provide high osmotic pressure, offer efficient recovery methods, have low toxicity, and deliver high water flux. DS can be separated into categories [63]:

- Inorganic salts
- Organic polymers
- Magnetic nanoparticles (MNP)
- Hydrogels
- Ionic liquids

Inorganic salts as DSs have been extensively researched and tested. The most commonly used inorganic salt DSs are NaCl, $NaNO_3$, KCl, KNO_3, NH_4Cl, NH_4NO_3, $CaCl_2$, and $Ca(NO_3)_2$ [33, 65]. Sodium chloride is the most commonly used DS since it is inexpensive, accessible, very soluble, and allows for high osmotic pressure [66]. Organic polymer DSs, such as glucose and fructose, do not require regeneration as the diluted DS created is typically drinkable or dischargeable. These DSs are not typically used in PW treatment. MNPs are a DS that has become inclined to many researchers due to their high recovery rate and osmotic pressure [63]. These DSs are created by using a polyol and thermal decomposition process to create hydrophilic magnetic nanoparticles [67]. The reverse salt flux of FO processes using MNPs as their draw solute is nearly nonexistent. The complication of MNP DSs is their cost. These DS are not only preparation intensive but also cost-intensive. MNP DS also has a propensity for fouling and lower water flux. Additionally, they decompose easily and can lead to the loss of draw solutes and the contamination of the newly recovered freshwater [67]. Hydrogels and ionic liquids are relatively recent additions to the studies on DSs. Hydrogels can provide high osmotic pressure, depending on the ionic group of the hydrogel polymer.

An essential part of the selection of DS is the regeneration method that is needed. These methods are thermal separation, membrane separation, thermal/membrane hybrid separation, stimuli-responsive separation, and chemical precipitation. If these methods are not utilized or required, the dilute DS is either disposed of or useable for drinking, fertilization, etc. [16]. In the case of PW treatment, the feed solution is the PW itself. Thus, a DS is appropriate for the FO membrane process and application in question. DS must provide a strong driving force for mass transport and the energy consumption associated with reconcentrating the draw solution for continuous FO operation [50].

4.2 FO membrane fouling

Fouling occurs when particles/contaminants are accumulated or absorbed by the membrane. This phenomenon can negatively affect the permeability, selectivity, and lifespan of the membrane, thus increasing the maintenance requirements and the cost. There are the main factors that affect membrane fouling: the properties of FS and the operating conditions. Membrane foulants can be separated into three categories: inorganic, organic, and biological. Fouling in membranes treating PW is specifically caused by the accumulation or absorption of oil droplets or solutes on the surface and the internal pore structure of the membrane. Fouling caused by PW can become extremely complex, with the possibility of all three foulants happening simultaneously. In a recent study, thin-film HF membranes were used to treat PW from unconventional oil and gas fields [68]. Several trials using this membrane, using two different PW samples and in the FO and PRO modes, were completed and compared. The results showed that when the membrane was in the PRO mode, it had a high propensity for fouling. It also showed that the faster the crossflow velocity, the more likely foulant build-up would occur. Comparatively, CTA membranes are more resistant to fouling than TFC membranes when treating PW [17]. Because fouling is generally unavoidable, strategies for cleaning the membranes and fouling mitigation must be employed. The most common methods employed are pretreatment, physical cleansing, chemical cleansing, and incorporating nanomaterials into the membrane, as discussed earlier. Most membranes require water flushing followed by chemical cleansing [17, 63].

4.3 FO membrane hybrid and integrated processes

To further improve efficiency, FO membranes that are useful on their own can be paired with other treatment methods. A widely studied hybrid FO process is FO integrated with RO. A FO-RO hybrid system is where one process acts as the treatment method and the other acts as the posttreatment method. The combination of the two membrane processes leads to over a 70% increase in potential water recovery [38]. This same method can be done with NF, MF, MD, OED, MED, MD, and ED. FO-NF is a more economically feasible hybrid method than FO-RO [69]. ED-FO hybrid processes, also known as ED-FO renewable energy desalination (EDFORD), differ from the other hybrid approaches in the matter that the system is powered by solar energy. Comparatively, the cost of this process was higher than the traditional FO membrane processes [15]. All hybrid FO membrane processes are similarly effective in treating FS.

5. Environmental impacts and economic analysis

Water is an invaluable resource vital for society and humanity to survive, grow, and thrive. From clothes, food, and personal hygiene to transportation, energy, and sanitation, these means are crucial to day-to-day life, and water is needed for each of these processes to be executed successfully. Energy production uses large amounts of water and makes up 75% of all industrial water usage [70]. With water and energy demand on the rise and 39% of the population globally being unable to access water that has been adequately treated, other methods of water treatment must be considered. This study has focused on water treatment as it applies to PW. The management and treatment of PW from oil and gas sites are costly. Many materials and resources go into the production of oil and gas, and it creates many environmental impacts. The various methods utilized to treat PW were discussed earlier. These methods, though plentiful, can be quite costly and labor-intensive. This is why new methods are being explored.

When considering a product or service, it is imperative to consider the impacts the product or service may have, environmentally and economically. One way to do this is to perform a life cycle assessment (LCA). Most LCAs include the following elements [71]:

- Extraction and preparation of raw materials
- Manufacturing
- Distribution
- Usage
- Repair, upgrade, and maintenance
- Final disposal or recycling

These elements consider the various impacts and effects that the product or service may have on the environment or the economy.

5.1 Life cycle assessment of FO for PW treatment

Life cycle assessment is a method used to evaluate and estimate the environmental impacts of a product or service. It considers the phases that the products may go through and the impacts that coincide with them. The goal of LCA is to compile all important information regarding the materials, energy, and environmental inputs and outputs, to interpret the results of the assessment, and to utilize them for application. The LCA method has four steps:

- Definition of scope & goal
- Analysis of inventory
- Assessment of impact
- Interpretation

The first three steps are done in succession, with the fourth being an assessment of the first three. The first step, defining the scope & goal, depends on the context and circumstances of the product or service. In this case, the product/service is FO membranes as they relate to PW treatment.

The following step is the analysis of inventory, where an account of the input and output completed. This account includes the materials and energy consumed and the emissions and waste generated during the life cycle of the product. Once this detailed account of the inputs and outputs is completed, it is labeled as the life cycle inventory (LCI). The next step is to assess the impact the product/service may have on the environment. There are four steps to this assessment, though only two are required. The first is classification, where the inputs and outputs analyzed beforehand are placed into categories defined in the first LCA step. The next is characterization, where the impact of each induvial category is evaluated. Subsequently to characterization is normalization, where the results of the previous are normalized using a reference factor to clarify and simplify the results. Lastly is weighting, where all indicators and results are translated to a singular index or score [71]. The last step of the LCA is interpretation, which is the finalization of any data, results, assessment, or analysis made during the three previous steps. This step aims to for a clear presentation of the entire LCA, giving clear and concise conclusions and assessments of the scope and goal of the product/service.

Martins et al. performed an LCA on various membrane processes. They found that there is a limited amount of materials on the application of membrane processes [71]. To complete the LCA method explicitly, more studies on membrane processes need to be completed. They suggested that these future studies include more details on the manufacture and preparation, maintenance, disposal, recycling, and sustainability of these membranes [71]. There are a limited number of studies discussing the social and economic impacts of membrane processes. The studies that include a discussion on these impacts do not coincide with the LCA method. The study does follow the LCA steps and uses the studies on the membranes that are available. The studies available were mostly membrane processes adopted for domestic consumption. Most of the water sources being treated were non-industrial, and were either freshwater, groundwater, or wastewater. This is not a broad enough spectrum to complete an LCA of the membranes, especially in regard to PW.

Caballero et al. completed utilizing the LCA method, though with different indicators and scales. They found that multi-effect evaporation with mechanical vapor recompression (MEE-MVR), was the best alternative to standard wastewater treatment in regards to the treatment of PW, with an environmental impact 21.9% lower than the single-effect evaporation with mechanical vapor recompression (SEE-MVR) technology [72]. The categories of their LCI that were most affected by the PW treatment were climate change and fossil depletion due to the use of electricity, stem, and evaporation technologies, depending on the method. Coday et al. also completed an LCA of PW, but with a much tighter scope. They chose to focus on Haynesville shale pit water [73]. Their study tested three different treatment methods and completed an LCA on each. These were compared to other treatment methods. The first method did not employ membrane technology, the second employed CTA FO, TFC RO, and TFC NF membranes with a NaCl DS, and the third employing osmotic dilution and the same CTA FO membrane from the second method. This study found the second scenario had the most negative impact on the environment [73].

5.2 Economic analysis of FO for PW treatment

Life cycle cost analysis (LCCA) is a method used to estimate the cost of a product or service for conception to disposal. Coday et al. also performed a cost analysis of the studies done on the Haynesville shale pit water. They concluded that employing FO membranes into treatment could be more cost-efficient when compared to the current disposal practice at Haynesville. Reuse and recycling of the water could benefit their economy by up to 60% [73]. This sentiment is echoed in Echchelh et al.'s analysis of the reuse/recycling of PW versus the current disposal method. Their study found that the cost of reusing the PW instead of disposing of it was much lower [74]. The main part of this cost variance is that the PW has to be transported and discharge, which can both be cost-intensive depending on the location of the oil or gas field. This money can instead be used to treat the PW, and the treated PW can be utilized for other purposes, such as agricultural irrigation. Other studies modeled the reused of PW in an oil/gas field, though they did not include an assessment in regards to FO membranes or the reuse of PW [72]. Another study did a cost analysis of three hybrid FO systems: FO-RO, FO-NF90, and FO-NF270 [69]. They compared the total water, chemical cleaning, DS regeneration, and overall annual cost of each FO system. All three systems had a higher overall cost compared to a seawater RO system, 37.5% higher on average [69]. This increase in cost was caused by the number of FO membrane modules required to treat the water. Though there was not a large price variation from system to system, price variation could be seen clearly in the DS used for each system. $MgCl_2$ systems were presented with the highest annual cost; thus, due to all the systems having similar annual cost, $MgCl_2$ is the cause of the spike in cost. Earlier, the importance of DS selection was discussed. This study echoes that sentiment. The cost of this DS resulted in a much higher annual cost, 67.0% higher than the system using NaCl as their DS [69].

6. Conclusion and future perspectives

Produced water treatment is a complex and costly process. This is due to the many characteristics and contaminants in PW. Current methods used to treat PW tend to be complex, costly, and harmful to the environment. New, more efficient methods are being considered. This study focuses on the application of FO membrane processes to treat PW. As discussed, FO has a myriad of characteristics and parameters that can be tailored to meet the requirements of PW to be treated. Though these membranes have an abundance of potential, there are still challenges these membranes must face. FO membranes are still being studied and tested. Most novel FO membranes discussed in this study are not able to be applied on a larger scale yet. There is also still the issue of finding a DS for PW water treatment that does not cost and maintenance intensive. FO membranes can also be integrated and combined with other membrane processes. These methods are currently more applicable due to these hybrid processes not requiring regeneration and having considerably lower fouling propensity. Fouling can also be mitigated by increasing or including a pretreatment process. There are currently not enough studies and applications of FO membranes in PW facilities and processes, making it complicated to perform a complete LCA and/or LCCA.

In the future, more studies where FO membranes and hybrid FO membrane systems are applied and tested using various DS should be completed. Furthermore, more of these studies should discuss the cost and environmental analysis of the membrane create. Currently, most studies include most of the formed and little to none of the latter. It's important to take into consideration not just how efficient a new treatment method is, but if efficiency is worth the cost in the long run. Currently, the LCA and LCCA done on membrane processes do show that FO membrane processes, hybrid, integrated, or not, are currently a feasible option on a large scale. Forward osmosis membranes can be competitive and applied widely but require additional research in membrane fabrication, better DS, fouling mitigation, and cleansing and application methods.

Acknowledgements

This work was primarily funded through a grant from NSF, Award #1900787. Partial support was also received from NSF CREST Center for Energy & Environmental Sustainability at Prairie View A&M University, NSF Award #1914692.

Author details

Layla Ogletree, Hongbo Du and Raghava Kommalapati*
Prairie View A&M University, Prairie View, United States

*Address all correspondence to: rrkommalapati@pvamu.edu

IntechOpen

© 2021 The Author(s). Licensee IntechOpen. This chapter is distributed under the terms of the Creative Commons Attribution License (http://creativecommons.org/licenses/by/3.0), which permits unrestricted use, distribution, and reproduction in any medium, provided the original work is properly cited.

References

[1] Igunnu ET, Chen GZ. Produced water treatment technologies. *Int J Low-Carbon Technol* 2014; 9: 157-177.

[2] Rangarajan T, Heydari A, Henni A. State of the Art Treatment of Produced Water. In: *Water Treatment*. InTech, 2013. Epub ahead of print January 16, 2013. DOI: 10.5772/53478.

[3] Scanlon BR, Reedy RC, Xu P, et al. Can we beneficially reuse produced water from oil and gas extraction in the U.S.? *Sci Total Environ*; 717. Epub ahead of print May 15, 2020. DOI: 10.1016/j.scitotenv.2020.137085.

[4] Fakhru'l-Razi A, Pendashteh A, Abdullah LC, et al. Review of technologies for oil and gas produced water treatment. *Journal of Hazardous Materials* 2009; 170: 530-551.

[5] Veil JA. *US produced water volumes and management practices*, http://www.circleofblue.org/waternews/wp-content/uploads/2010/09/ANL_EVS__R09_produced_water_volume_report_2437.pdf (2015, accessed November 16, 2020).

[6] Fillo JP, Evans JM. Characterization and management of produced waters from underground natural gas storage reservoirs. In: *American Gas Association Operation Section Proceedings*. 1990, pp. 448-459.

[7] United States Environmental Protection Agency (USEPA). Profile of the Oil and Gas Extraction Industry. *EPA Off Compliance Sect Noteb Proj*, http://www.epa.gov/compliance/resources/publications/assistance/sectors/notebooks/oilgas.pdf (2000).

[8] Tibbetts PJC, Buchanan IT, Gawel LJ, et al. A comprehensive determination of produced water composition. In: Ray JP, Engelhardt FR (eds) *Produced Water: Technological/Environmental Issues and Solutions*. New York: Plenum Publishing Corp., 1992, pp. 97-113.

[9] Shepherd MC, Shore FL, Mertens SK, et al. Characterization of produced waters from natural gas production and storage operations: regulatory analysis of a complex matrix. In: Ray JP, Engelhardt FR (Eds. . (eds) *Produced Water: Technological/Environmental Issues and Solutions*. New York: Plenum Publishing Corp., 1992, pp. 163-174.

[10] Stephenson M. A survey of produced water studies. In: Ray J, Engelhardt F (eds). (eds) *Produced Water: Technological/Environmental Issues and Solutions*. Plenum Publishing Corp., 1992, pp. 1-12.

[11] Boukhriss M, Zhani K, Ben Bacha H. Optimization of membrane distillation (MD) technology for specific application desalination. *Int J Adv Manuf Technol* 2017; 88: 55-66.

[12] Munirasu S, Haija MA, Banat F. Use of membrane technology for oil field and refinery produced water treatment - A review. *Process Saf Environ Prot* 2016; 100: 183-202.

[13] Coday BD, Cath TY. Forward osmosis: Novel desalination of produced water and fracturing flowback. *J Am Water Works Assoc* 2014; 106: 37-38.

[14] Coday BD, Xu P, Beaudry EG, et al. The sweet spot of forward osmosis: Treatment of produced water, drilling wastewater, and other complex and difficult liquid streams. *Desalination* 2014; 333: 23-35.

[15] Akther N, Sodiq A, Giwa A, et al. Recent advancements in forward osmosis desalination: A review. *Chem Eng J* 2015; 281: 502-522.

[16] Francis L, Ogunbiyi O, Saththasivam J, et al. A comprehensive

review of forward osmosis and niche applications. In: *Environmental Science: Water Research and Technology*. Royal Society of Chemistry, 2020. Epub ahead of print August 1, 2020. DOI: 10.1039/d0ew00181c.

[17] Ahmad NA, Goh PS, Yogarathinam LT, et al. Current advances in membrane technologies for produced water desalination. *Desalination* 2020; 493: 114643.

[18] Shin S, Kim AS. Temperature Effect on Forward Osmosis. In: Du H, Thompson A, Wang X (eds) *Osmotically Driven Membrane Processes - Approach, Development, and Current Status*. IntechOpen, 2018. Epub ahead of print 2018. DOI: 10.5772/intechopen.72044.

[19] Eyvaz M, Arslan S, İmer D, et al. Forward Osmosis Membranes – A Review: Part I. 2016; 22.

[20] Tian M, Wang YN, Wang R. Synthesis and characterization of novel high-performance thin-film nanocomposite (TFN) FO membranes with nanofibrous substrate reinforced by functionalized carbon nanotubes. *Desalination* 2015; 370: 79-86.

[21] Wang KY, Chung T, Amy G. Developing thin-film-composite forward osmosis membranes on the PES/SPSf substrate through interfacial polymerization. *AIChE J* 2011; 58: 770-781.

[22] Wang KY, Ong RC, Chung TS. Double-skinned forward osmosis membranes for reducing internal concentration polarization within the porous sublayer. *Ind Eng Chem Res* 2010; 49: 4824-4831.

[23] Tang YP, Luo L, Thong Z, et al. Recent advances in membrane materials and technologies for boron removal. *Journal of Membrane Science* 2017; 541: 434-446.

[24] Qin JJ, Lay WCL, Kekre KA. Recent developments and future challenges of forward osmosis for desalination: A review. *Desalin Water Treat* 2012; 39: 123-136.

[25] HTI Water. Forward Osmosis Water Filtration Products, http://www.htiwater.com/ (accessed November 21, 2020).

[26] McCutcheon JR, McGinnis RL, Elimelech M. A novel ammonia—carbon dioxide forward (direct) osmosis desalination process. *Desalination* 2005; 174: 1-11.

[27] Suwaileh W, Pathak N, Shon H, et al. Forward osmosis membranes and processes: A comprehensive review of research trends and future outlook. *Desalination* 2020; 485: 114455.

[28] Mohammadifakhr M, Grooth J de, Roesink HDW, et al. Forward osmosis: A critical review. *Processes* 2020; 8: 1-29.

[29] Lee KP, Arnot TC, Mattia D. A review of reverse osmosis membrane materials for desalination-Development to date and future potential. *J Memb Sci* 2011; 370: 1-22.

[30] Gadallah H, Ali HM, Ali SS, et al. Application of Forward/Reverse Osmosis Hybrid System for Brackish Water Desalination using El-Salam Canal Water, Sinai, Egypt, Part (1): FO Performance. *4th Int Conf Environ Sci Eng* 2014; 68: 6-18.

[31] Lutchmiah K, Verliefde ARD, Roest K, et al. Forward osmosis for application in wastewater treatment: A review. *Water Res* 2014; 58: 179-197.

[32] McGovern RK, Lienhard V JH. On the potential of forward osmosis to energetically outperform reverse osmosis desalination. *J Memb Sci* 2014; 469: 245-250.

[33] Zhao D, Chen S, Guo CX, et al. Multi-functional forward osmosis

draw solutes for seawater desalination. *Chinese J Chem Eng* 2016; 24: 23-30.

[34] Xie M, Lee J, Nghiem LD, et al. Role of pressure in organic fouling in forward osmosis and reverse osmosis. *J Memb Sci* 2015; 493: 748-754.

[35] Lee S, Boo C, Elimelech M, et al. Comparison of fouling behavior in forward osmosis (FO) and reverse osmosis (RO). *J Memb Sci* 2010; 365: 34-39.

[36] Bhinder A, Shabani S, Sadrzadeh M. Effect of Internal and External Concentration Polarizations on the Performance of Forward Osmosis Process. In: Du H, Thompson A, Wang X (eds) *Osmotically Driven Membrane Processes*. IntechOpen, 2018. Epub ahead of print 2018. DOI: 10.5772/intechopen.71343.

[37] She Q, Wang R, Fane AG, et al. Membrane fouling in osmotically driven membrane processes: A review. *Journal of Membrane Science* 2016; 499: 201-233.

[38] Mahto A, Aruchamy K, Meena R, et al. Forward osmosis for industrial effluents treatment – sustainability considerations. *Sep Purif Technol*; 254. Epub ahead of print 2021. DOI: 10.1016/j.seppur.2020.117568.

[39] Alsvik IL, Hägg MB. Pressure retarded osmosis and forward osmosis membranes: Materials and methods. *Polymers (Basel)* 2013; 5: 303-327.

[40] Li Y, Pan G, Wang J, et al. Tailoring the Polyamide Active Layer of Thin-Film Composite Forward Osmosis Membranes with Combined Cosolvents during Interfacial Polymerization. *Ind Eng Chem Res* 2020; 59: 8230-8242.

[41] Loeb S, Titelman L, Korngold E, et al. Effect of porous support fabric on osmosis through a Loeb-Sourirajan type asymmetric membrane. *J Memb Sci* 1997; 129: 243-249.

[42] Mulder M. *Basic principles of membrane technology*. Dordrecht, The Netherlands: Kluwer Academic Publishers, 1992.

[43] Aende A, Gardy J, Hassanpour A. Seawater desalination: A review of forward osmosis technique, its challenges, and future prospects. *Processes*; 8. Epub ahead of print 2020. DOI: 10.3390/PR8080901.

[44] Ghaffour N, Lattemann S, Missimer T, et al. Renewable energy-driven innovative energy-efficient desalination technologies. *Appl Energy* 2014; 136: 1155-1165.

[45] Khaydarov RA, Khaydarov RR. Solar-powered direct osmosis desalination. *Desalination* 2007; 217: 225-232.

[46] Saren Q, Qiu CQ, Tang CY. Synthesis and characterization of novel forward osmosis membranes based on layer-by-layer assembly. *Environ Sci Technol* 2011; 45: 5201-5208.

[47] Hołda AK, Vankelecom IFJ. Understanding and guiding the phase inversion process for synthesis of solvent resistant nanofiltration membranes. *J Appl Polym Sci* 2015; 132: 1-17.

[48] Liang S, Gao P, Gao X, et al. Improved blending strategy for membrane modification by virtue of surface segregation using surface-tailored amphiphilic nanoparticles. *Front Environ Sci Eng* 2016; 10: 1-9.

[49] Li L, Wu J, Gao C. Surface-grafted block copolymer brushes with continuous composition gradients of poly(poly(ethylene glycol)-monomethacrylate) and poly(N-isopropylacrylamide). *Sci China Chem* 2011; 54: 334-342.

[50] Klaysom C, Cath TY, Depuydt T, et al. Forward and pressure retarded

osmosis: Potential solutions for global challenges in energy and water supply. *Chem Soc Rev* 2013; 42: 6959-6989.

[51] Cath TY, Childress AE, Elimelech M. Forward osmosis: Principles, applications, and recent developments. *Journal of Membrane Science* 2006; 281: 70-87.

[52] Suzaimi ND, Goh PS, Ismail AF, et al. Strategies in forward osmosis membrane substrate fabrication and modification: A review. *Membranes (Basel)* 2020; 10: 1-42.

[53] Tang Y, Li S, Xu J, et al. Thin-film composite forward osmosis membrane with single-walled carbon nanotubes interlayer for alleviating internal concentration polarization. *Polymers (Basel)*; 12. Epub ahead of print 2020. DOI: 10.3390/polym12020260.

[54] Zhao D, Qiu G, Li X, et al. Zwitterions coated hollow fiber membranes with enhanced antifouling properties for osmotic power generation from municipal wastewater. *Water Res* 2016; 104: 389-396.

[55] Shokrollahzadeh S, Tajik S. Fabrication of thin-film composite forward osmosis membrane using electrospun polysulfone/polyacrylonitrile blend nanofibers as porous substrate. *Desalination* 2018; 425: 68-76.

[56] Lau WJ, Gray S, Matsuura T, et al. A review on polyamide thin-film nanocomposite (TFN) membranes: History, applications, challenges, and approaches. *Water Res* 2015; 80: 306-324.

[57] Liu Q, Qiu G, Zhou Z, et al. An Effective Design of Electrically Conducting Thin-Film Composite (TFC) Membranes for Bio and Organic Fouling Control in Forward Osmosis (FO). *Environ Sci Technol* 2016; 50: 10596-10605.

[58] Han C, Zhang X, Ding C, et al. Improved performance of thin-film composite membrane supported by aligned nanofibers substrate with slit-shape pores for forward osmosis. *J Memb Sci*; 612. Epub ahead of print 2020. DOI: 10.1016/j.memsci.2020.118447.

[59] Chiao YH, Chen ST, Patra T, et al. Zwitterionic forward osmosis membrane modified by fast second interfacial polymerization with enhanced antifouling and antimicrobial properties for produced water pretreatment. *Desalination* 2019; 469: 114090.

[60] Wu W, Shi Y, Liu G, et al. Recent development of graphene oxide-based forward osmosis membrane for water treatment: A critical review. *Desalination* 2020; 491: 114452.

[61] Zhong P, Fu X, Chung TS, et al. Development of thin-film composite forward osmosis hollow fiber membranes using direct sulfonated polyphenylenesulfone (sPPSU) as membrane substrates. *Environ Sci Technol* 2013; 47: 7430-7436.

[62] Arkhangelsky E, Wicaksana F, Tang C, et al. Combined organic-inorganic fouling of forward osmosis hollow fiber membranes. *Water Res* 2012; 46: 6329-6338.

[63] Li L, Shi W, Yu S. Research on forward osmosis membrane technology still needs improvement in water recovery and wastewater treatment. *Water (Switzerland)* 2020; 12: 1-27.

[64] Tang CY, Zhao Y, Wang R, et al. Desalination by biomimetic aquaporin membranes: Review of status and prospects. *Desalination* 2013; 308: 34-40.

[65] Chaoui I, Abderafi S, Vaudreuil S, et al. Water desalination by forward osmosis: draw solutes and recovery

methods–review. *Environ Technol Rev* 2019; 8: 25-46.

[66] Zeng L-M, Du M-Y, Wang X-L. A Thermodynamical Approach for Evaluating Energy Consumption of the Forward Osmosis Process Using Various Draw Solutes. *Water* 2017; 9: 189.

[67] Mishra T, Ramola S, Shankhwar AK, et al. Use of synthesized hydrophilic magnetic nanoparticles (HMNPs) in forward osmosis for water reuse. *Water Sci Technol Water Supply* 2016; 16: 229-236.

[68] Liden T, Carlton DD, Miyazaki S, et al. Comparison of the degree of fouling at various flux rates and modes of operation using forward osmosis for remediation of produced water from unconventional oil and gas development. *Sci Total Environ* 2019; 675: 73-80.

[69] Kim JE, Phuntsho S, Chekli L, et al. Environmental and economic assessment of hybrid FO-RO/NF system with selected inorganic draw solutes for the treatment of mine impaired water. *Desalination* 2018; 429: 96-104.

[70] UNESCO. *Water and Climate Change: Facts and Figures*. 2020.

[71] Martins AA, Caetano NS, Mata TM. LCA for Membrane Processes. *Sustain Membr Technol Water Wastewater Treat* 2017; 23-66.

[72] Caballero JA, Labarta JA, Quirante N, et al. Environmental and economic water management in shale gas extraction. *Sustain*; 12. Epub ahead of print 2020. DOI: 10.3390/su12041686.

[73] Coday BD, Miller-Robbie L, Beaudry EG, et al. Life cycle and economic assessments of engineered osmosis and osmotic dilution for desalination of Haynesville shale pit water. *Desalination* 2015; 369: 188-200.

[74] Echchelh A, Hess T, Sakrabani R. Agro-environmental sustainability and financial cost of reusing gasfield-produced water for agricultural irrigation. *Agric Water Manag* 2020; 227: 105860.

Section 3

Advancements in Osmotically Driven Membrane Systems

Chapter 3

Forward Osmosis Membrane Technology in Wastewater Treatment

Deniz Şahin

Abstract

In recent times, membrane technology has proven to be a more favorable option in wastewater treatment processes. Membrane technologies are more advantageous than conventional technologies such as efficiency, space requirements, energy, quality of permeate, and technical skills requirements. The forward osmosis (FO) membrane process has been widely applied as one of the promising technologies in water and wastewater treatment. Forward osmosis uses the osmotic pressure difference induced by the solute concentration difference between the feed and draw solutions. The proces requires a semi-permeable membrane which has comparable rejection range in size of pollutants (1 nm and below). This chapter reviews the application of FO membrane process in wastewater treatment. It considers the advantages and the disadvantages of this process.

Keywords: Desalination, Forward osmosis, FO-Based Hybrid System, Integrated FO System, Wastewater, Wastewater treatment

1. Introduction

Membrane separation processes are widely used in the last decade for industrial, commercial, and domestic activities such as water and wastewater treatment, energy-efficiency. Within the concentration-driven processes, FO has gained increasing prominence due to its advantages such as possibility of low fouling, high salt rejection, and high water recovery. However, FO does have inherently disadvantages such as; reverse solute diffusion (RSD), lower flux, concentration polarization (CP), and membrane fouling. These obstacles oblige the developing new processes, synthesis of different membrane materials or modifications, and finding new draw solution (DS). There is therefore an exigent need to develop new FO membranes by optimization of thickness, porosity, tortuosity of active/support layer of FO membrane.

This chapter is divided into two parts. In this first part of chapter, basic principles of FO phenomenon, advantages and challenges of FO over conventional membrane processes are addressed by the literature review and scholarly articles. The second part of which states applications of FO process for wastewater remediation, and recent developments in FO process.

2. General aspects of forward osmosis

2.1 Process description

Forward osmosis is one example of water separation processes and a potential acceptable alternative/complement to reverse osmosis (RO) process for power generation, wastewater treatment and desalination. Forward osmosis is a membrane process in which requires little or no hydraulic pressure. Unlike the RO process, in the FO process, an osmotic pressure gradient through a semi-permeable membrane is the driving force of water transport from the feed solution (FS) to the DS [1]. Thus, the concentrated DS generates an osmotic pressure and drives water from the feed through the membrane while most of the contaminants and salts are rejected by the membrane, then separating the water from the diluted DS [2]. **Figure 1** illustrates the principle of operation of RO and FO processes.

The general equation used to describe theoretical water flux across the RO and FO membrane (*Jw*) is calculated using Darcy's law [1]:

$$Jw = Aw \times (\sigma \Delta \pi - \Delta P) \quad (1)$$

where, *Aw* is the membrane pure water permeability coefficient, σ is the reflection coefficient which indicates the rejection capability of a membrane (for a perfect semipermeable membrane $\sigma = 1$), $\Delta\pi$ is the osmotic pressure differential across the membrane, and ΔP is the applied external pressure. Therefore, in FO, ΔP is zero thus making the water flux to be directly proportional to the difference in osmotic pressure, while for RO, $\Delta P > \Delta\pi$. The relation between water flux and applied pressure is illustrated in **Figure 2**.

2.2 Draw solution

Both FO and RO processes involve semi-permeable membranes as key component, which has comparable rejection range in size of pollutants (1 nm and below). One of the major factors in the development of FO membrane is selecting an appropriate DS [4]. The ideal DS should have following characteristics: high osmotic pressure, low molecular weight (MW), non-toxicity, relatively low-cost, high water solubility, and efficiently regeneration [5, 6]. Sodium chloride (NaCl)

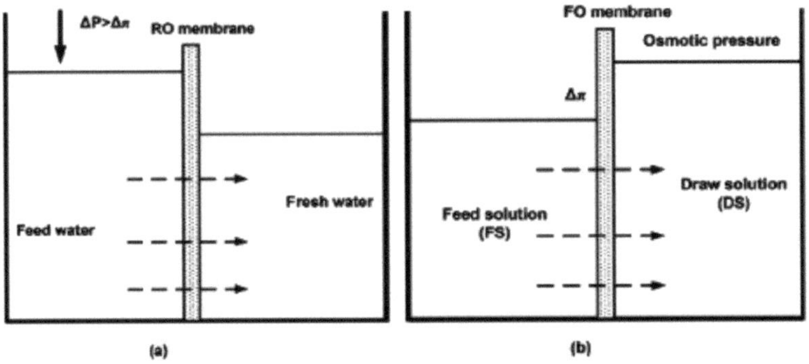

Figure 1.
Schematic illustration of the (a) RO, and (b) FO processes.

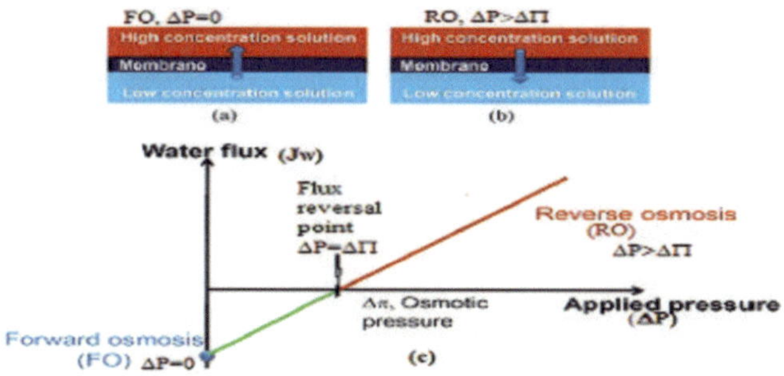

Figure 2.
Schematic representation of FO, RO process: (a) FO process where no external force is applied on the high concentration solution. The natural flow of water is from the low concentration side to the high concentration side; and (b) RO process where applied pressure on the high concentration solution exceeds the osmotic pressure difference across the membrane, so the water flux is opposite to the flux in FO process; and (c) classification of these processes in a water flux vs. applied pressure. Adapted from [3, 4].

is among the most commonly used draw solute in FO because it has highly water solubility and it is also relatively easy to reconcentrate using classical desalination processes [1]. In the past few decades, vast studies have been performed to determine desirable DSs, the different DSs are presented in **Table 1**, such as (1) inorganic compounds (e.g., NaCl, sodium nitrate ($NaNO_3$), magnesium sulfate ($MgSO_4$)) (2), organic compounds (e.g., glucose, fructose, 2-methylimidazole-based compounds) (3), functionalized nanoparticles (e.g., magnetic nanoparticles (MNPs)), Na^+-functionalized carbon quantum dots (Na-CQD).

The different DSs allow the generate of high osmotic pressure and can be easily regenerated or recovered. Nevertheless, their costs have not been successfully determined [39].

2.3 Membrane material

The identification of an ideal membrane in FO process is a key component which needs to be addressed to further advance this process. A perfect semipermeable membrane should have high water flux and solutes rejection, low propensity to fouling, and high chemical and thermal stability and so forth [2].

The FO membrane can be either synthetic or natural. In the early studies, the variety of natural materials used has included animal bladders and intestines [4]. A few decades ago, investigators have been examined different materials for FO membrane fabrication that include cellulose, rubber, and porcelain [4, 40, 41]. Although synthetic FO membranes have been currently commercially available; but this technology is still in its infancy. As a result, many types of FO membranes have been investigated that are able to perform well under a very wide range of applications [42–51]. **Table 2** provides information about membranes used in wastewater treatment.

As can be seen from **Table 2**, CTA-FO membranes have been used in the most of the experimental working on wastewater treatment due to its relatively higher tolerance to chlorine, insensitive to bio-degradation, and low fouling potential [66–68]. Despite its advantages, there are still some drawbacks such as narrow pH range, relatively low water permeability and high NaCl permeability [69–71]. Compared with CTA membranes, TFC membranes have higher fouling propensity, higher surface selectivity, a wider pH range, and better chemical stability [72–75]. Although CTA membranes have also a chlorine tolerance of up to 1 ppm (part per

Categories	Draw Solutes	Recovery Methods	Ref.
Inorganic compounds	NaCl	reverse osmosis (RO)	[7, 8]
	inorganic fertilizer	direct use	[9, 10]
	potassium sulfate (K_2SO_4)	RO	[7]
	sodium nitrate ($NaNO_3$)	direct use	[10]
	aluminum sulfate ($Al_2(SO_4)_3$)	precipitation	[11]
	magnesium sulfate ($MgSO_4$), copper sulfate ($CuSO_4$)	precipitation	[12, 13]
Organic compounds	glucose, fructose, sucrose	RO	[12, 14–17]
	2-Methylimidazole compounds	membrane distillation (MD)	[18]
	sodium polyacrylate (PAA-Na)	ultrafiltration (UF), MD	[19, 20]
	poly (aspartic acid sodium salt)	MD	[21]
	N,N-dimethylcyclohexylamine (N(Me)2Cy)	heating	[22]
	1--Cyclohexylpiperidine (CHP)	heating	[23]
	Micellar solution	UF	[24]
	oxalic acid complexes with Fe/Cr/Na	nanofiltration (NF)	[25]
	trimethylamine–carbon dioxide	heating	[26]
	CO2-responsive polymers (PDMAEMA)	UF	[27]
	poly(sodium styrene-4-sulfonate-co-N-isopropylacrylamide) (PSSS-PNIPAM)	MD	[28]
	Switchable polarity solvent (SPS)	RO	[29]
	polyelectrolyte incorporated with triton-x114	MD	[30]
	dimethyl ether	heating with solar energy	[31]
	poly(4-styrenesulfonic acid-co-maleic acid)	NF	[32]
Functional nanoparticles	Super hydrophilic nanoparticles	UF	[33]
	hydrophilic superparamagnetic nanoparticles	magnetic separation	[34]
	magnetic core-hydrophilic shell nanosphere	magnetic separation	[35]
	thermoresponsive Magnetic Nanoparticle	magnetic separation	[36]
	dextran-coated MNPs magnetic separation	magnetic separation	[37]
	hyperbranched polyglycerol coated MNPs	magnetic separation	[38]

Table 1.
Overview of the different DSs in FO process.

million), TFC membranes have limited tolerance to chlorine attack [76]. On the other hand, TFC membranes prone to membrane fouling which negatively impacts their operational and maintenance costs.

Feed	Membrane	Findings	Ref.
Wastewater containing heavy metals	Lab scale (thin film composite) TFC membrane	Synthetic, good flux in PRO mode only	[52]
Synthetic dye wastewater	Cellulose-acetate (CA) hollow fiber Lab Scale	High viscosity, synthetic.	[53]
Wastewater with sludge	Cellulose tri-acetate (CTA)-HTI	Phosphorous recovery from sludge.	[54]
Polyvinyl chloride (PVC) latex	CTA-HTI	No regeneration required.	[55]
Synthetic wastewater	CTA-HTI	No regeneration.	[56]
Synthetic wastewater	Flat sheet biomimetic membrane by aquaporin A/S	Microbial cells in DS can lead to biofouling. No regeneration required.	[56]
Biorefineries	Flat sheet biomimetic membrane by aquaporin A/S	DS can be toxic. No regeneration.	[57]
Textile wastewater	Biomimetic aquaporin A/S	High RSF for dye mixtures. No regeneration is required in case of dye mixture DS.	[58]
Printed circuit board (PCB) plant wastewater	TFC porifera	DS leads to inorganic scaling. No regeneration required.	[59]
Medical radioactive liquid wastewater	TFC polyamide (PA) membrane porifer	NaCl has a higher rejection for Iodine.	[60]
Synthetic wastewater & municipal treatment plans wastewater	CTA-HTI	Same flux for FO and FO_wEO (electrochemical oxidation).	[61]
Seawater	CTA-HTI	Feed flow rate of 2.9 L/min, No space and pretreatment.	[62]
Oily wastewater	Lab scale TFC-polyethersulfone (PES) membrane	In PRO mode oxalic acid had good flux.	[63]
Oily wastewater	TFC Cellulose acetat butyrate(CAB) holow fiber Lab scale	The experiment was done in the PRO mode. This membrane had excellent oil rejection.	[64]
Synthetic wastewater	TFC-ES HTI	Presence of cations in feed aggravates fouling in FO.	[65]

Table 2.
Some previous and recent researches on FO membranes.

In addition to fouling of membrane, concentration polarization has an impact on the water flux, particularly at the support layer, which leads to the severity in internal concentration polarization (ICP). A low ICP requires a low S-value (structural parameter) [43, 77].

The membrane structural parameter S is defined as [2]:

$$S = KxD = \frac{ts \times Ƭ}{\varepsilon} \tag{2}$$

where *D* is the diffusion coefficient of the draw solute, *ts* is the thickness of the support layer, Ƭ is the tortuosity, and ε is the porosity of the support layer.

Recently, new materials have been investigated for FO membrane fabrication to increase water flux, reduce ICP, and enhance the tolerance to water quality.

3. Application in wastewater treatment

As an emerging membrane technology, FO has been investigated over the last decade for seawater or brackish water desalination, wastewater treatment, power generation, pharmaceutical applications, and food&dairy processing in both academic research and industries [78, 79].

The most attractive usage of FO is its application for wastewater treatment. Consequently, there are two clusters of applications (i) desalination and (ii) water reuse (**Figure 3**) [80].

Key attributes of this process are:

- high rejection of a wide range of contaminants,
- lower energy consumption,
- high water recovery,
- lower brine discharge,
- lower membrane fouling propensity.

However, the main challenges in this process are related to:

- Development of high performance, such as higher permeate water flux and lower reverse salt flux of FO membranes,
- Reducing concentration polarization in membranes,
- Ensuring low DS reverse solute diffusion through the membrane,
- Adaptive reuse
- Regeneration of the DS.

3.1 Desalination

Saline water (e.g. seawater or brackish water) and an osmotic reagent (e.g. a non-volatile or a volatile salt) are used as the FS and DS, respectively, in the *direct FO desalination* [81, 82]. In this process, after the FO process, an additional step is needed to recycle the draw solutes as well as to produce purified water [83, 84]. One of the first examples of FO application in water desalination was published in 1975. This study was intended to desalinate Atlantic Ocean seawater to produce an emergency water supply on lifeboats by *direct* osmosis (**Figure 4**) across a CA-FO membrane with a hypertonic glucose solution as the DS [85]. In another study, a flat-sheet CTA-FO membrane was used in seawater desalination, yielding a high water flux and high salt rejection (over 95%) with 6 M ammonium bicarbonate (NH_4HCO_3) as DS [84]. Also, polymer hydrogels particles have been studied as draw agents in FO desalination. Smaller polymer hydrogel particles led to higher FO water flux in these tests. Similarly, higher salt concentration led to lower FO

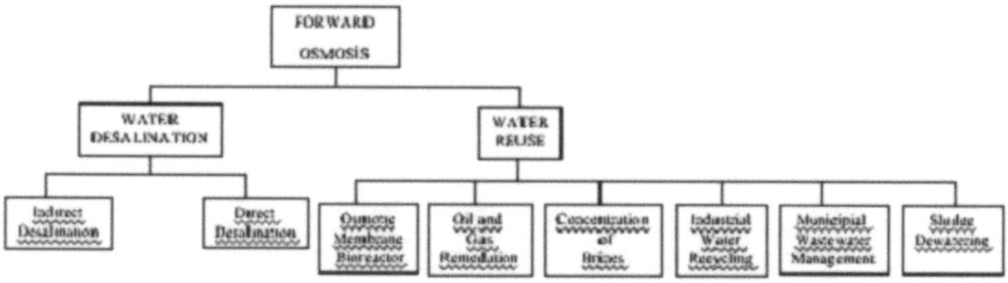

Figure 3.
Applications of FO in the water treatment industry.

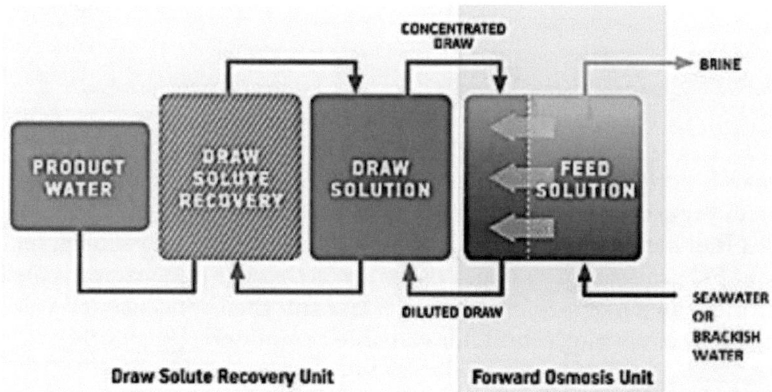

Figure 4.
FO process for desalination of seawater or brackish water.

water flux. Meanwhile, the use of a commercial FO membrane was more suitable than RO membrane [83]. Another study modified magnetic particles covered with thermo sensitive polymer investigated as DS and about 93% of salt recovery was obtained [34]. The world's first commercial forward osmosis desalination plant for direct sea water treatment was established in Al Najdah, Oman. This facility is still in operation and has reduced chemical consumption and provides higher throughput and longer membrane life, significant operational and capital costs and to be more reliable than traditional methods [86]. Membrane fouling and scaling problems at RO stage mitigate due to the use of FO as a pretreatment step for the RO process.

Indirect FO desalination uses a high salinity water (e.g. seawater or brackish water) as a natural DS and quality-impaired water source (e.g. wastewater effluent or urban storm water runoff) as the feed solution [87, 88]. The diluted seawater or brackish water can potentially couple with low pressure reverse osmosis (LPRO). The FO-LPRO hybrid process has lower costs for producing water compared to pure reverse osmosis [89]. These experiments have demonstrated the ability of FO membranes to reject nutrients from wastewater, especially chemical oxygen demand (COD) and phosphate, and moderately nitrogen compounds [88, 90]. As an example, a submerged membrane module which makes it possible to adapt the process to a primary clarifier tank has been employed for partial desalination of seawater. The findings indicated that FO membranes have high rejection of heavy metals present in the wastewater (~99%). This study also showed that the use of biopolymers-like substances resulted in the fouling layer on the membrane surface [88]. A similar result has been reported in the use of osmotic membrane bioreactor (OMBR) for municipal wastewater treatment [91].

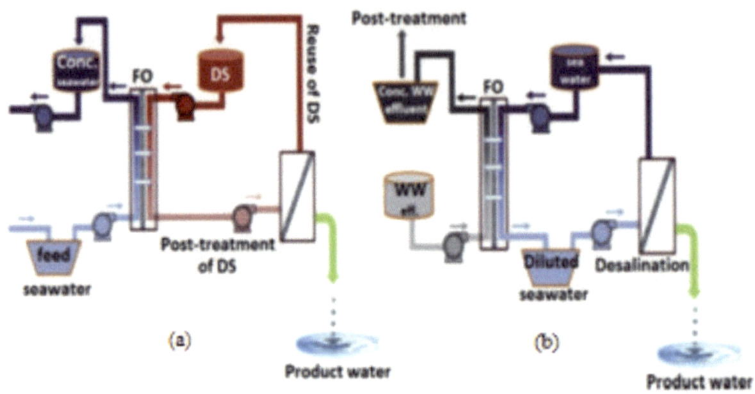

Figure 5.
Scheme of the two FO processes for desalination (a) direct, (b) indirect (adapted from [92]).

Direct and indirect arrangements of desalination systems using FO membrane are shown in **Figure 5**.

On the other hand, the pretreatment of wastewater has not yet been reported in the study of FO process. The reason, probably, is that the FO system is considered as a pre-treatment step to concentrate wastewater and then concentrated wastewater can be used to recover biogas or other valuable compounds [88, 93, 94].

3.2 Wastewater treatment

Forward osmosis has been utilized to treat various types of wastewater such as municipal wastewater (sewage) [95–98], oily wastewater [67, 99, 100], tanner effluent [101], automobile effluents [102], dairy streams [102, 103], produced water [104–106] besides synthetic wastewater [107, 108].

Lately, the current systems on FO application on wastewater treatment may be classified into two groups: FO and FO-based hybrid processes, and integrated FO processes. Both in FO and FO-based hybrid systems, the FO membrane is used to recover fresh water and reject of pollutants from the feed solution. In the integrated FO system, the FO membrane gradually replaces conventional membrane in the bioreactor, such as the FO membrane in membrane bioreactor (MBR). The function of the membrane is to concentrate the wastewater and improve the performance of the modified system.

Therefore, FO has been extensively applied in wastewater treatment and reuse, resource recovery, seawater desalination, and food/medicine manufacturing as shown in **Table 3**.

The FO process shows promising results for the treatment of wastewater, and has many advantages in comparison to the conventional wastewater treatment processes. When high process recoveries are obtained, FO processes become viable. Forward osmosis also provides a more sustainable flux and reliable removal of contaminants.

3.2.1 FO and FO-based hybrid system

Hybrid desalination systems using emerging FO process and combined with traditional process like reverse osmosis, membrane distillation, nanofiltration, electrodyalsis (ED) could potentially reduce the energy consumption of the desalination process, and decrease obstacles in the implementation of process. In these systems, FO is used as a pre-treatment step, while RO, NF, and ED are

Field		FS	DS	Process	Ref.
Wastewater treatment and reuse	Tannery wastewater	Tannery wastewater	NaCl solution	FO	[101]
	High-salinity oil-bearing wastewater	Oil-bearing wastewater	3 M NaCl	FO	[109]
	Oil sands tailings water	OSPW	Basal depressurization water	FO	[110]
Resource Recovery	P and N recovery from urine	Fresh urine	Mg-based fertilizer DS	FO	[111]
	Precious metal recovery	Pd ion waste solution	Electroless (E'less) nickel (Ni) waste solution	FO	[112]
	Energy recovery	algae culture wastewater	Seawater	FO	[113]
Seawater desalination	/	Wastewater	Seawater	FO + DS	[62]
Food and medicine manufacturing	/	Sugarcane juice	Sea bittern	FO	[114]
	/	Protein	Superabsorbent polymer(SAP) hydrogels	FO	[115]
	/	Medical radioactive liquid waste	NaCl	FO	[60]

Table 3.
Application of FO in different industries.

known as water recovery or draw solution regeneration/reconcentration step [116, 117]. An overview of FO and FO-based hybrid system configurations is depicted in **Table 4**.

3.2.1.1 Hybrid FO-MD system

The performance of the FO process can be improved by its combination with other system to take advantage of the unique strengths of the individual processes. For this reason, FO process is often combined with an MD process (**Figure 6**). As an example, the FO-MD hybrid system was employed for raw sewage [93] at water recovery up to 80%. This process also achieved high removal efficiency for trace organic contaminants (TrOCs) that rates 91–98%. In another study, this hybrid system was used for oily wastewater treatment. The findings indicated that 90% feed water recovery could be readily attained with trace amounts of oil and NaCl [99]. A vapor pressure driving FO-MD system was studied for treatment high salinity hazardous waste landfill leachate [129]. Total organic carbon (TOC) and total nitrogen (TN) rejection rates were higher than 98% while rejection rate of salt was higher than 96%. NH_4^+-N, and heavy metal ions were also completely removed. Similar performance could also be seen in the application of dairy wastewater and grain possessing wastewater treatment [103, 130].

Hybrid System	DS	FS	Ref.
FO	Fertilizer chemicals	Municipal wastewater	[9]
FO	3 M NaCl	Oil-bearing wastewater	[109]
FO	Basal depressurization water	OSPW	[110]
FO	10% NaCl	Coal gasification wastewater	[118]
FO-MD	$MgCl_2$	Digested sludge	[94]
	NaCl	Oily wastewater	[99]
	NaCl	Salinity landfill leachate	[129]
	$MgSO_4$	Dairy and grain wastewater	[130]
FO-RO	NaCl, Na_2SO_4, $MgSO_4$	Synthetic feed (NaCl); groundwater (Mawson Lakes, South Australia)	[116]
	NaCl, $MgCl_2$	Seawater (TDS = 32000–45000 mg/L)	[119]
	Red Sea seawater (TDS = 40.5 g/L)	Wastewater effluent (Al Ruwais wastewater treatment plant, Jeddah, Saudi Arabia)	[120]
	Seawater after UF	Coal-fired power plant wastewater	[121]
FO-NF	Na_2SO_4	Brackish water from Mawson Lakes, South Australia (TDS = 3970 mg/L)	[122]
	NaCl, KCl, $CaCl_2$, $MgCl_2$, $MgSO_4$, Na_2SO_4 and $C_6H_{12}O_6$	Simulated seawater (0.6 M NaCl)	[123]
	NaCl, $CaCl_2$, $MgSO_4$, Na_2SO_4	A site located in northwest Italy	[124]
	$MgCl_2$	Municipal wastewater	[108]
FDFO + ED	1 M DAP	Treated wastewater (secondary effluent)	[125]
FO + ED-RO	NaCl	Seawater	[126]
FO + ED	/	brackish and wastewater	[127]
FO + ED-RO	/	Seawater	[128]

Table 4.
An overview of FO and FO-based hybrid systems.

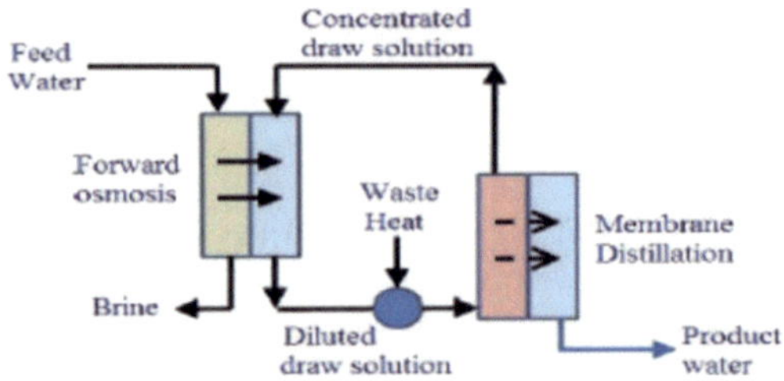

Figure 6.
Schematic diagram of hybrid system consisting of FO and MD processes.

3.2.1.2 Hybrid FO-RO system

Due to the current scenario of global water crisis, seawater desalination has become one of the practical solutions to produce water of potable quality. Membrane based desalination processes have been used to desalinate seawater have been widely reported. Among the various desalination processes, RO is the most consistent and reliable process which offers a number of advantages due to its high salt rejection rate, high quality drinking water, high water recovery, and green technology [131]. Despite the aforementioned advantages, several shortcomings, such as high energy consumption and severe fouling propensity remain the obstacles [132]. In recent years, the hybrid system of the FO and RO processes has gained increasing prominence among researchers [8, 116, 117, 119]. As can be seen in **Figure 7**, the hybrid system consists of two stages. The first stage begins with the migration of fresh water from the seawater feed solution to join the draw solution. In the second stage, the product fresh water is separated from the draw solution in the RO unit [89].

In the first study focusing on this FO-RO hybrid system, the authors demonstrated that the approach may provide four major benefits over stand-alone RO desalination: lower energy use, multi-barrier protection of drinking water, beneficial reuse of impaired water, reduction in RO membrane fouling [89]. Similar interest has also been conducted that compaires the hybrid FO-RO system and the stand-alone RO process for seawater desalination [119]. The study showed that the hybrid FO-RO system can be highly competitive depending on the salinity of seawater and type and concentration of the draw solute. Interestingly, total power consumption in a hybrid FO-RO system was higher than that in RO process, yet the FO process alone was only contributed 2–4% of the total power consumption in the FO-RO hybrid system. Therefore, most of the power consumption in the FO-RO system was realized in the high hydraulic pressure RO regeneration unit [119]. In another study, FO process used as a pre-treatment for a hybrid FO-RO desalination system. The optimal parameters such as water flux, water recovery and final draw solution of this FO pre-treatment process were determined by modeling and were experimentally validated by using real brackish water [116]. In a further study, FO-RO hybrid system for coal-fired power plant wastewater treatment, seawater after UF was investigated as DS. Results showed that the total energy consumption of the FO-RO system was 15% less than that of a typical seawater desalination RO [121].

3.2.1.3 Hybrid FO-NF system

The literature includes theoretical studies on the strengthening economic and environmental potential of the large-scale FO-based systems but very few experimental reports exist on these issues [133–135]. Examples include discussion on pilot-scale FO coupled with NF and other distillation processes for treating wastewater effluents. For example; a pilot-scale FO-NF hybrid closed loop system was developed for the treatment of tannery wastewater at a rate of 52–55 L/m^{2}h and rejections of 98.5% COD, 97.2% chlorides and 98.2% sulfate were achieved [136].

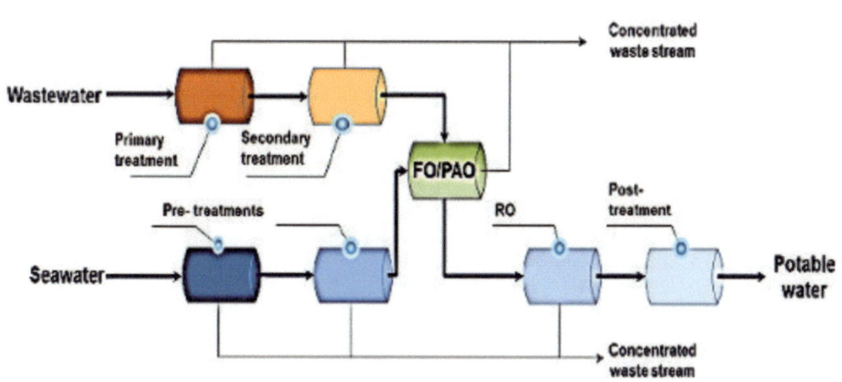

Figure 7.
Schematic diagram of the hybrid FO-RO system (adapted from [89]).

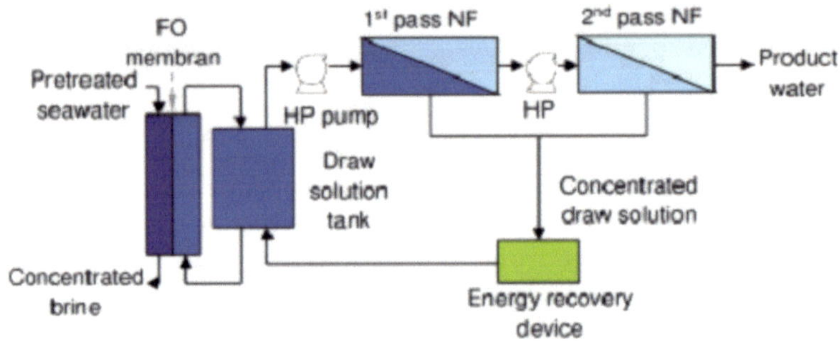

Figure 8.
Schematic diagram of the hybrid FO-NF system for seawater desalination (adapted from [137]).

In addition, a hybrid FO–NF system designed for brackish water desalination was investigated and also presented promising results such as lower hydraulic pressure, less flux decline [122]. In another study, a hybrid FO-NF system with two NF passes for the post treatment was used for desalinating seawater [123]. A proposed configuration of a hybrid FO-NF process for seawater desalination is shown in **Figure 8** [137].

3.2.1.4 Hybrid FO-ED System

Electrodyalsis is a membrane-based separation process in which ions across ion-selective membranes under an electric field. A FO-ED hybrid system was investigated by using diammonium phosphate (DAP), as DS to achieve wastewater reuse and mitigation of salinity buildup on the feed side. Electrodyalsis was able to significantly recover the 96.6 ± 3.0% reverse-fluxed DAP under 3.0 V 1-h daily operation [125]. Forward osmosis process was tested upstream to ED-RO system for an access to DS with higher electrical conductivity in the FO-ED-RO hybrid system [126]. In another study, FO-ED-RO hybrid system proposed to produce high-quality water from secondary-effluent or brackish water is shown in **Figure 9**. Results showed that the water from this system contains a low concentration of total organic carbon (TOC), carbonate and cations derived from the feed water [127].

3.2.2 Integrated FO system

The integrated FO system includes an osmotic microbial fuel cell (OsMFC) and osmotic membrane bioreactor (OMBR). Recent research has elucidated how

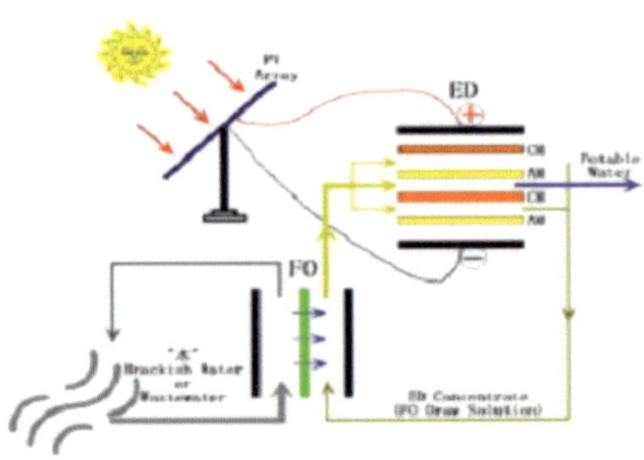

Figure 9.
Schematic diagram of a novel photovoltaic powered FO-ED system (adapted from [127]).

the integration of osmosis in MFC and MBR was used through the application of FO membrane for simultaneous recovery of osmotic water, the concentration of wastewater, and the improvement of effluent quality [138, 139].

3.2.2.1 OsMFC

The system uses FO integrated into a microbial fuel cell (MFC) to improve the quality of the treated wastewater and the performance of the fuel cell. A FO membrane is placed between the anode chamber with wastewater and the cathode chamber full of DS and water flux through this membrane transports protons from the anode to the cathode [140–145]. An OsMFC (**Figure 8**) achieved water flux of 3.94 ± 0.22 L/m^2h with a catholyte containing 2 M NaCl, while there was no obvious water flux in a conventional MFC [140]. In a further study, FO membrane is integrated into an air-cathode MFC (AAFO-MFC) for enhancing bio-electricity and water recovery from low-strength wastewater. The AAFO-MFC system produced a high quality effluent, with the removal rates of organic contaminants and total phosphorus (P) of more than 97% [145].

There are also some drawbacks for OsMFC application in wastewater treatment such as the lower water flux of the FO membrane, membrane fouling and salt accumulation (**Figure 10**) [146].

3.2.2.2 OMBR

Hollow fiber or flat-sheet MF and UF membranes are commonly used membranes in MBR. A major problem associated with the operation of MF-UF-MBRs is membrane fouling. A novel MBR-named OMBR- has been developed and widely used to reduce fouling and promote the reuse of treated wastewater. In OMBR, FO membrane module is displaced in the wastewater. A combination biological treatment and an OMBR uses to remove water from the mixed liquor to the draw side under the osmotic pressure gradient. The pollutants, activated sludge and solids are all rejected by the membrane. The OMBR-based hybrid system, for the first time, was utilized to direct recovery nutrient from municipal wastewater with over 90% of nutrient. In this study, nutrient and mineral salts were rejected via FO membrane and enriched within the bioreactor and then recovered by chemical precipitation [147]. The OMBR has several advantages, including higher rejection rate, lower energy consumption, and higher quality of treated wastewater compared to the traditional MBR. However, OMBR still has some disadvantages, such as salinity accumulation and membrane fouling. Based on the OMBR hybrid system, an integrated UF or MF membrane system in the OMBR system was investigated to remove the soluble inorganic salts in the reactor [148]. This process has a longer sludge

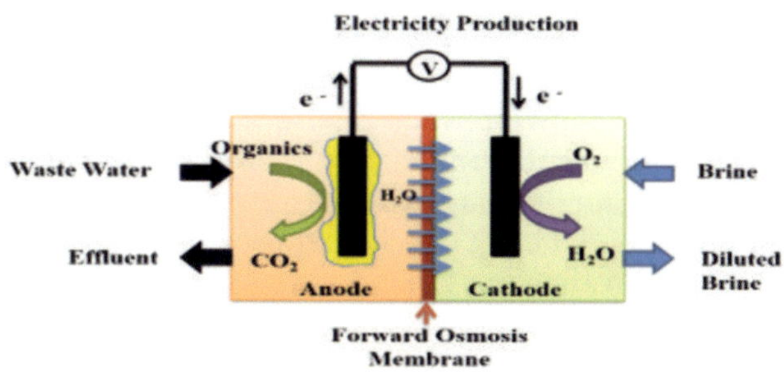

Figure 10.
Schematic diagram of an OsMFC (adapted from [140]).

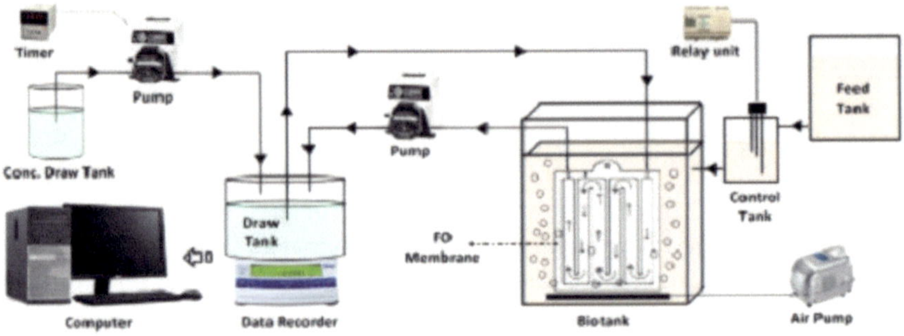

Figure 11.
Schematic diagram of an OMBR (adapted from [150]).

residence time (SRT) than the traditional OMBR system, so a higher sludge concentration can be obtained. Similarly, MF membrane was added to the system for phosphate recovery from the raw sewage, in which MF and FO membranes function in parallel. The results show that the phosphate can be recycled up to 98%. The MF membrane retained phosphate and mineral salts in the bioreactor, so phosphate was precipitated as calcium phosphate precipitates without the input of Ca^{2+} ions [149]. In another study, the OMBR system was operated in treating of Chromium (Cr) and Lead (Pb) metals of the high strength wastewater. The findings revealed that industrial wastewater containing more than 5 mg/L of Cr and more than 2 mg/L of Pb is not recommended for the OMBR due to poor sludge characteristics, and high membrane fouling (**Figure 11**) [150].

4. Conclusions

The FO membrane process is a promising process for drinking water purification and wastewater treatment technology due to its excellent high rejection rate performance and relatively low membrane fouling characteristics. Hence it is likely to gain an very important place in the membrane technology.

The engineering of the FO process application is relatively scarce, due to the FO investigations and applications are still in the laboratory scale and progress in practical applications still requires further proof of the pilot. The research on membrane fouling mechanism is also needed, which still has a large gap in the current research results. Over the past decade, a large number of research papers has been published on membrane development (to increase water flux) and process design (i.e., to increase osmotic pressure, to change sludge retention time) and the number of papers in these issues has also increased year by year. The researchers' focus is to develop next-generation membranes by advanced membrane fabrication methods as well as hybrid systems where the FO process can really add value.

This chapter focuses mainly on forward osmosis either individually or in combination with other processes for wastewater treatment. For example; the FO removes the large molecular weight trace organic compounds while the combination of the MBR and NF/RO process for removing TrOCs from synthetic wastewater is feasible. The key concepts mentioned in the chapter provide better understanding for further promoting the utilization of FO process and its new applications for water resource recovery and wastewater treatment development.

Author details

Deniz Şahin
Faculty of Science, Department of Chemistry, Gazi University, Ankara, Turkey

*Address all correspondence to: dennoka1k@hotmail.com

IntechOpen

© 2021 The Author(s). Licensee IntechOpen. This chapter is distributed under the terms of the Creative Commons Attribution License (http://creativecommons.org/licenses/by/3.0), which permits unrestricted use, distribution, and reproduction in any medium, provided the original work is properly cited.

References

[1] Jasmina, K., Subhankar, B., Malini, B., Claus, H-N., Irena, P., Forward osmosis in wastewater treatment processes. Acta Chimica Slovenica, 2017, 64 (1), 83-94.

[2] Li, D., Yan, Y., Wang, H., Recent advances in polymer and polymer composite membranes for reverse and forward osmosis processes. Progress in Polymer Science, 2016, 61, 104-155.

[3] Chun, Y., Mulcahy, D., Zou L., Kim I.S., A short review of membrane fouling in forward osmosis processes. Membranes, 2017, 7(30), 1-23.

[4] Cath, T.Y., Childress, A.E., Elimelech, M., Forward osmosis: Principles, applications, and recent developments. J. Membr. Sci., 2006, 281, 70-87.

[5] Yong, J.S., Phillip, W.A., Elimelech, M., Reverse permeation of weak electrolyte draw solutes in forward osmosis. Ind. Eng. Chem. Res., 2012, 51, 13463-13472.

[6] Chung, T.-S., Zhang, S., Wang, K.Y., Su, J., Ling, M.M., Forward osmosis processes: Yesterday, today and tomorrow. Desalination, 2012, 287, 78-81.

[7] Ansari A.J., Faisal I.H., Wenshan G., Hao, H.N., William. E.P., Hai Long. N.D., Factors governing the pre-concentration of wastewater using forward osmosis for subsequent resource recovery. Sci. Total Environ., 2016, 566-567, 559-566.

[8] Achilli, A., Cath, T.Y., Childress, A.E., Selection of inorganic-based draw solutions for forward osmosis applications. J. Membr. Sci., 2010, 364, 233-241.

[9] Chekli, L., Kim, Y., Phuntsho, S., Li, S., Ghaffour, N., Leiknes, T., Shon, H.K., Evaluation of fertilizer- drawn forward osmosis for sustainable agriculture and water reuse in arid regions. J. Environ.Manag., 2017, 187, 137-145.

[10] Phuntsho, S., Shon, H.K., Hong, S., Lee, S., Vigneswaran, S.A., Novel low energy fertilizer driven forward osmosis desalination for direct fertigation: Evaluating the performance of fertilizer draw solutions. J. Membr. Sci., 2011, 375, 172-181.

[11] Frank, B.S., Desalination of Sea Water. SU.S. Patent US 3670897A, 20 June 1972.

[12] Garcia-Castello, E.M., McCutcheon, J.R., Elimelech, M., Performance evaluation of sucrose concentration using forward osmosis. J. Membr. Sci. 2009, 338, 61-66.

[13] Alnaizy, R., Aidan, A., Qasim, M. Draw solute recovery by metathesis precipitation in forward osmosis desalination. Desalin. Water Treat. 2013, 51, 5516-5525.

[14] Tan, C.H., Ng, H.Y., Revised external and internal concentration polarization models to improve flux prediction in forward osmosis process. Desalination, 2013, 309, 125-140.

[15] Kravath, R.E., Davis, J.A., Desalination of sea water by direct osmosis. Desalination, 1975, 16, 151-155.

[16] Stache, K., Apparatus for Transforming Sea Water, Brackish Water, Polluted Water or the Like into a Nutrious Drink by Means of Osmosis. U.S. Patent US 4879030A, 7 November 1989.

[17] Yaeli, J., Method and apparatus for processing liquid solutions of suspensions particularly useful in the desalination of saline water. U.S. Patent US 5098575A, 24 March 1992.

[18] Yen, S.K., Haja, M., N.F., Su, M., Wang, K.Y., Chung, T.-S., Study of draw solutes using 2-methylimidazole-based compounds in forward osmosis. J. Membr. Sci. 2010, 364, 242-252.

[19] Ge, Q., Su, J., Amy, G.L., Chung, T.-S., Exploration of polyelectrolytes as draw solutes in forward osmosis processes. Water Res. 2012, 46, 1318-1326.

[20] Ge, Q., Wang, P., Wan, C., Chung, T.-S., Polyelectrolyte-promoted forward osmosis-membrane distillation (FO-MD) hybrid process for dye wastewater treatment. Environ. Sci. Technol., 2012, 46, 6236-6243.

[21] Gwak, G., Jung, B., Han, S., Hong, S., Evaluation of poly (aspartic acid sodium salt) as a draw solute for forward osmosis. Water Res., 2015, 80, 294-305.

[22] Reimund, K.K., Coscia, B.J., Arena, J.T., Wilson, A.D., McCutcheon, J.R., Characterization and membrane stability study for the switchable polarity solvent N,N-dimethylcyclohexylamine as a draw solute in forward osmosis. J. Membr. Sci., 2016, 501, 93-99.

[23] Orme, C.J., Wilson, A.D., 1-cyclohexylpiperidine as a thermolytic draw solute for osmotically driven membrane processes. Desalination, 2015, 371, 126-133.

[24] Roach, J.D., Abdulrahman, A.-A., Alaa, A.-N., Mohammed, H., Use of micellar solutions as draw agents in forward osmosis. J. Surfactants Deterg., 2014, 17, 1241-1248.

[25] Ge, Q., Fu, F., Chung, T.-S., Ferric and cobaltous hydroacid complexes for forward osmosis (FO) processes. Water Res., 2014, 58, 230-238.

[26] Boo, C., Khalil, Y.F., Elimelech, M., Performance evaluation of trimethylamine–carbon dioxide thermolytic draw solution for engineered osmosis. J. Membr. Sci., 2015, 473, 302-309.

[27] Tian, E., Hu, C., Qin, Y., Ren, Y., Wang, X., Wang, X., Xiao, P., Yang, X., A study of poly (sodium 4-styrenesulfonate) as draw solute in forward osmosis. Desalination, 2015, 360, 130-137.

[28] Zhao, D., Wang, P., Zhao, Q., Chen, N., Lu, X., Thermoresponsive copolymer-based draw solution for seawater desalination in a combined process of forward osmosis and membrane distillation. Desalination, 2014, 348, 26-32.

[29] Stone, M.L., Rae, C., Stewart, F.F., Wilson, A.D., Switchable polarity solvents as draw solutes for forward osmosis. Desalination, 2013, 312, 124-129.

[30] Ray, S.S., Chen, S.S., Nguyen, N.C., Nguyen, H.T., Dan, N.P., Thanh, B.X., Trang, L.T., Exploration of polyelectrolyte incorporatedwith Triton-X 114 surfactant based osmotic agent for forward osmosis desalination. J. Environ. Manag., 2018, 209, 346-353.

[31] Monjezi, A.A., Mahood, H.B., Campbell, A.N., Regeneration of dimethyl ether as a draw solute in forward osmosis by utilising thermal energy from a solar pond. Desalination, 2017, 415, 104-114.

[32] Huang, J., Long, Q., Xiong, S., Shen, L., Wang, Y., Application of poly(4-styrenesulfonic acid-co-maleic acid) sodium salt as novel draw solute in forward osmosis for dye-containing wastewater treatment. Desalination, 2017, 421, 40-46.

[33] Ling, M.M., Chung, T.-S., Desalination process using super hydrophilic nanoparticles via forward

osmosis integrated with ultrafiltration regeneration. Desalination, 2011, 278, 194-202.

[34] Li, Z., Wei, L., Gao, M.Y., Lei, H., One-pot reaction to synthesize biocompatible magnetite nanoparticles. Adv. Mater., 2005, 17, 1001-1005.

[35] Park, S.Y., Ahn, H.-W., Chung, J.W., Kwak, S.-Y., Magnetic core-hydrophilic shell nanosphere as stability-enhanced draw solute for forward osmosis (FO) application. Desalination, 2016, 397, 22-29.

[36] Zhao, Q., Chen, N., Zhao, D., Lu, X., Thermoresponsive magnetic nanoparticles for seawater desalination. ACS Appl. Mater. Interfaces, 2013, 5, 11453-11461.

[37] Bai, H., Liu, Z., Sun, D.D., Highly water soluble and recovered dextran coated Fe_3O_4 magnetic nanoparticles for brackish water desalination. Sep. Purif. Technol., 2011, 81, 392-399.

[38] Hee-Man, Y., Bum-Kyoung, S.E.O., Kune-Woo, L.E.E., Jei-Kwon, M., Hyperbranched polyglycerol-coated magnetic nanoparticles as draw solute in forward osmosis. Asian J. Chem., 2014, 26, 4031-4034.

[39] Boo, C., Elimelech, M., Hong, S., Fouling control in a forward osmosis process integrating seawater desalination and wastewater reclamation. J. Membr. Sci., 2013, 444, 148-156.

[40] Zhang, J., Forward osmosis membrane bioreactor for water reuse in Civil and Environmental Engineering. 2011, National university of Singapore: Singapore. p. 120.

[41] Anderson, D.K., Concentration of dilute industrial wastes by direct osmosis. 1977: University of Rhode Island. 364.

[42] Yang, Q., Wang K.Y., and Chung T.-S., Dual-layer hollow fibers with enhanced flux as novel forward osmosis membranes for water production. Environ. Sci. Technol., 2009, 43(8), 2800-2805.

[43] Yip, N.Y., Tiraferri A., Phillip W.A, Schiffman J.D., Elimelech M., High performance thin-film composite forward osmosis membrane. Environ. Sci. Technol, 2010, 44(10), 3812-3818.

[44] Wei, J., Qiu, C., Tang C.Y., Wang, R., Fane, A.G., Synthesis and characterization of flat-sheet thin film composite forward osmosis membranes. J. Membr. Sci., 2011, 372(1-2), 292-302.

[45] Qiu, C., Setiawan, L., Wang, R., Tang C.Y., Fane, A.G., High performance flat sheet forward osmosis membrane with an NF-like selective layer on a woven fabric embedded substrate. Desalination, 2011, 287, 266-270.

[46] Ong, R.C. and Chung T.-S., Fabrication and positron annihilation spectroscopy (PAS) characterization of cellulose triacetate membranes for forward osmosis. J. Membr. Sci., 2012, 394-395, 230-240.

[47] Wang, K.Y., Ong R.C., and Chung T.-S., Double-skinned forward osmosis membranes for reducing internal concentration polarization within the porous sublayer. Industrial & Engineering Chemistry Research, 2010, 49(10), 4824-4831.

[48] Tiraferri, A., Yip, N.Y., Phillip W.A, Schiffman J.D., Elimelech M., Relating performance of thin-film composite forward osmosis membranes to support layer formation and structure. J. Membr. Sci., 2011, 367(1-2), 340-352.

[49] Wang, K.Y., Chung T.-S., and Amy G., Developing thin-film-composite forward osmosis membranes on the PES/SPSf substrate through

interfacial polymerization. AIChE Journal, 2011, 58(3), 770-781.

[50] Wang, K.Y., Yang, Q., Chung T.-S., Rajagopalan R., Enhanced forward osmosis from chemically modifiedpolybenzimidazole (PBI) nanofiltration hollow fiber membranes with a thin wall. Chemical Engineering Science, 2009, 64(7), 1577-1584.

[51] Zhang, S., Wang, K.Y., Chung, T.-S., Chen, H., Jean, Y.C., Amy, G., Well-constructed cellulose acetate membranes for forward osmosis: Minimized internal concentration polarization with an ultra-thin selective layer. J. Membr. Sci., 2010, 360(1-2), 522-535.

[52] Cui, Y., Ge, Q., Liu, X.-Y., and Chung, T.-S., Novel forward osmosis process to effectively remove heavy metal ions. J. Membr. Sci., 2014, 467, 188-194.

[53] Ge, Q., Wang, P., Wan, C., and Chung, T.-S., Polyelectrolyte-promoted forward osmosis–membrane distillation (FO–MD) hybrid process for dye wastewater treatment. Environ. Sci. Technol., 2012, 46, 6236-6243.

[54] Ansari, A.J., Hai, F.I., Price, W. E., and Nghiem, L.D., Phosphorus recovery from digested sludge centrate using seawater-driven forward osmosis. Separation and Purification Technology, 2016, 163, 1-7.

[55] Takahashi, T., Yasukawa, M., and Matsuyama, H. Highly condensed polyvinyl chloride latex production by forward osmosis: Performance and characteristics. J. Membr. Sci., 2016, 514, 547-555.

[56] Zou, S., and He, Z., Enhancing wastewater reuse by forward osmosis with self-diluted commercial fertilizers as draw solutes. Water Research, 2016, 99, 235-243.

[57] Kalafatakis, S., Braekevelt, S., Carlsen, V., Lange, L., Skiadas, I. V., and Gavala, H. N., On a novel strategy for water recovery and recirculation in biorefineries through application of forward osmosis membranes. Chemical Engineering Journal, 2017, 311, 209-216.

[58] Korenak, J., Helix-Nielsen, C., Buksek, H., and Petrinic, I., Efficiency and economic feasibility of forward osmosis in textile wastewater treatment. Journal of Cleaner Production, 2019, 210, 1483-1495.

[59] Gwak, G., Kim, D. I., and Hong, S., New industrial application of forward osmosis (FO): Precious metal recovery from printed circuit board (PCB) plant wastewater. J. Membr. Sci., 2018, 552, 234-242.

[60] Lee, S., Kim, Y., Park, J., Shon, H. K., and Hong, S., Treatment of medical radioactive liquid waste using Forward Osmosis (FO) membrane process. J. Membr. Sci., 2018, 556, 238-247.

[61] Liu, P., Zhang, H., Feng, Y., Shen, C., and Yang, F., Integrating electrochemical oxidation into forward osmosis process for removal of trace antibiotics in wastewater. Journal of Hazardous Materials, 2015, 296, 248-255.

[62] Hawari, A. H., Al-Qahoumi, A., Ltaief, A., Zaidi, S., and Altaee, A. Dilution of seawater using dewatered construction water in a hybrid forward osmosis system. Journal of Cleaner Production, 2018, 195, 365-373.

[63] Ge, Q., Amy, G. L., and Chung, T.-S., Forward osmosis for oily wastewater reclamation: Multi-charged oxalic acid complexes as draw solutes. Water Research, 2017, 122, 580-590.

[64] Han, G., de Wit, J. S., and Chung, T.-S., Water reclamation from emulsified oily wastewater via effective forward osmosis hollow fiber

membranes under the PRO mode. Water Research, 2015, 81, 54-63.

[65] Motsa, M.M., Mamba, B.B., and Verliefde, A.R.D., Forward osmosis membrane performance during simulated wastewater reclamation: Fouling mechanisms and fouling layer properties. Journal of Water Process Engineering, 2018, 23, 109-118.

[66] Herron, J. Asymmetric forward osmosis membranes: US patent. 2008. 7445712B2.

[67] Lv, L., Xu, J., Shan, B., Gao, C., Concentration performance and cleaning strategy for controlling membrane fouling during forward osmosis concentration of actual oily wastewater. J. Membr. Sci., 2017, 523, 15-23.

[68] Thorsen, T., Concentration polarisation by natural organic matter (NOM) in NF and UF. J. Membr. Sci., 2004, 233(1-2), 79-91.

[69] Chou, S., Shi, L., Wang, R., Tang, C. Y., Qiu, C., Fane, A. G., Characteristics and potential applications of a novel forward osmosis hollow fiber membrane. Desalination, 2010, 261(3), 365-372.

[70] Ren, J., and McCutcheon, J. R., A new commercial thin film composite membrane for forward osmosis. Desalination, 2014, 343, 187-193.

[71] Wang, C., Li, Y., Wang, Y., Treatment of greywater by forward osmosis technology: Role of the operating temperature. Environmental Technology, 2018, 1-10.

[72] Bell, E.A., Poynor, T.E., Newhart, K.B., Regnery, J., Coday, B.D., Cath, T.Y., Produced water treatment using forward osmosis membranes: Evaluation of extended-time performance and fouling. J. Membr. Sci., 2017, 525, 77-88.

[73] Chen, G., Wang, Z., Nghiem, L.D., Li, X., Xie, M., Zhao, B., Zhang, M., Song, J., He, T. Treatment of shale gas drilling flowback fluids (SGDFs) by forward osmosis: Membrane fouling and mitigation. Desalination, 2015, 366, 113-120.

[74] Coday, B.D., Almaraz, N., Cath, T.Y., Forward osmosis desalination of oil and gas wastewater: Impacts of membrane selection and operating conditions on process performance. J. Membr. Sci., 2015, 488, 40-55.

[75] Duong, P.H.H., Chung, T.-S., Application of thin film composite membranes with forward osmosis technology for the separation of emulsified oil-water. J. Membr. Sci., 2014, 452, 117-126.

[76] Fam, W., Phuntsho, S., Lee, J. H., Shon, H. K., Performance comparison of thin film composite forward osmosis membranes. Desalination and Water Treatment, 2013, 51(31-33), 6274-6280.

[77] Bui, N.-N., Arena, J. T., McCutcheon, J. R., Proper accounting of mass transfer resistances in forward osmosis: Improving the accuracy of model predictions of structural parameter. J. Membr. Sci., 2015, 492, 289-302.

[78] Korenak J., Basu S., Balakrishnan M., Hélix-Nielsen C., and Petrini I., Forward osmosis in wastewater treatment processes. Acta Chim. Slov., 2017, 64, 83-94.

[79] Lu, Y., Qin, M., Yuan, H., Abu-Reesh, I.M., He, Z., When bioelectrochemical systems meet forward osmosis:accomplishing wastewater treatment and reuse through synergy. Water, 2015, 7, 38-50.

[80] Zhao S., Zou L., Tang C. Y., Mulcahy D., Recent development in forward osmosis: opportunities and challenges. J. Membr. Sci., 2012, 396, 1-21.

[81] Chekli, L., Phuntsho, S., Shon, H.K., Vigneswaran, S., Kandasamy, J., Chanan, A., A review of draw solutes in forward osmosis process and their use in modern applications. Desalin. Water Treat., 2012, 43 (1-3), 167-184.

[82] Ge, Q., Ling, M., Chung, T.-S., Draw solutions for forward osmosis processes: developments, challenges, and prospects for the future. J. Membr. Sci. 2013, 442, 225-237.

[83] Li, D., Zhang, X., Simon, G.P., Wang, H., Forward osmosis desalination using polymer hydrogels as a draw agent: influence of draw agent, feed solution and membrane on process performance. Water Res., 2013, 47(1), 209-215.

[84] McCutcheon, J.R., McGinnis, R.L., Elimelech, M., A novel ammonia-carbon dioxide forward (direct) osmosis desalination process. Desalination, 2005, 174, 1-11.

[85] Kravath, R.E., Davis, J.A., Desalination of sea water by direct osmosis. Desalination, 1975, 16, 151-155.

[86] Modern Water. Membrane Processes Forward Osmosis: Desalination [Internet]. 2013. Available from: https://www.modernwater.com/pdf/MW_Factsheet_Membrane_HIGHRES.pdf [Accessed: Sep 14, 2017]

[87] Li, Z., Valladares Linares, R., Abu-Ghdaib, M., Zhan, T., Yangali-Quintanilla, V., Amy, G., Osmotically driven membrane process for the management of urban runoff in coastal regions.Water Res., 2014, 48, 200-209.

[88] Valladares Linares, R., Li, Z., Abu-Ghdaib, M., Wei, C.-H., Amy, G., Vrouwenvelder, J.S., Water harvesting from municipal wastewater via osmotic gradient: an evaluation of process performance. J. Membr. Sci., 2013, 447, 50-56.

[89] Cath, T.Y., Hancock, N.T., Lundin, C.D., Hoppe-Jones, C., Drewes, J.E., A multi-barrier osmotic dilution process for simultaneous desalination and purification of impaired water. J. Membr. Sci., 2010, 362, 417-426.

[90] Cath, T.Y., Drewes, J.E., Lundin, C.D., A novel hybrid forward osmosis process for drinking water augmentation using impaired water and Saline water sources. In: Proceedings of the 24th Annual WaterReuse Symposium, September 13-16, 2009, Seattle, Washington.

[91] Zhang, J., Loong, W.L.C., Chou, S., Tang, C., Wang, R., Fane, A.G., Membrane biofouling and scaling in forward osmosis membrane bioreactor. J. Membr. Sci., 2012, 403-404, 8-14.

[92] Li, Z., Valladares Linares, R., Muhannad, A., Amy, G., Comparative assessment of forward osmosis (FO) niches in desalination. In: Proceedings of IDA World Congress, October, 20-25, 2013, Tianjin, China.

[93] Xie, M., Nghiem, L.D., Price, W.E., Elimelech, M., A forward osmosis membrane distillation hybrid process for direct sewer mining: system performance and limitations. Environ. Sci. Technol., 2013, 47 (23), 13486-13493.

[94] Xie, M., Nghiem, L.D., Price, W.E., Elimelech, M., Toward resource recovery from wastewater: extraction of phosphorus from digested sludge using a hybrid forward osmosis membrane distillation process. Environ. Sci. Technol. Lett., 2014, 1 (2), 191-195.

[95] Sun Y., Tian J., Zhao Z., Shi W., Liu D., Cui F., Membrane fouling of forward osmosis (FO) membrane for municipal wastewater treatment: A comparison between direct FO and OMBR. Water Research, 2016, 104, 330-339.

[96] Han, G., Zhao, B., Fu, F., Chung, T-S., Weber, M., Staudt, C., Maletzko., High performance thin-film composite membranes with mesh-reinforced hydrophilic sulfonated polyphenylenesulfone (SPPSU) substrates for osmotically driven processes. J. Membr. Sci., 2016, 502, 84-93.

[97] Kim, S., Go, G-W., Jang, A., Study of flux decline and solute diffusion on an osmotically driven membrane process potentially applied to municipal wastewater reclamation. Journal of Industrial and Engineering Chemistry, 201, 33, 255-261.

[98] Bell, E.A., Holloway, R.W., Cath, T.Y., Evaluation of forward osmosis membrane performance and fouling during long-term osmotic membrane bioreactor study. J. Membr. Sci., 2016, 517, 1-13.

[99] Zhang, S., Wang, P., Fu X., Chung, T-S., Sustainable water recovery from oily wastewater via forward osmosis-membrane distillation (FO-MD). Water Research, 2014, 52, 112-121.

[100] Han, G., de Wit, J.S., Chung, T-S., Water reclamation from emulsified oily wastewater via effective forward osmosis hollow fiber membranes under the PRO mode. Water Research, 2015, 81, 54-63.

[101] Ahmed, F.A., Isam, I.O., Hasan, M., Forward osmosis process as an alternative method for the biological treatment of wastewater from the Al-Za'afaraniya tanning factory. The International Journal of Science and Technoledge, 2015, 3(1), 259-270.

[102] Haupt, A., Lerch, A., Forward osmosis treatment of effluents from dairy and automobile industry - Results from short-term experiments to show general applicability. Water Science and Technology, 2018, 78(3), 467-475.

[103] Song, H., Xie, F., Chen, W., Liu, J., FO/MD hybrid system for real dairy wastewater recycling. Environmental Technology, 2017, 39(18), 2411-2421.

[104] Bell, E.A., Poynor, T.E., Newhart, K.B., Regnery, J., Coday, B.D., Cath, T.Y., Produced water treatment using forward osmosis membranes: Evaluation of extended-time performance and fouling. J. Membr.Sci., 2016, 525, 77-88.

[105] Coday, B.D., and Cath, T.Y., Forward Osmosis: Novel desalination of produced water and fracturing flowback. Desalination, 2014, 106(2), 55-66.

[106] Liden, T., Hildenbrand, Z.L. and Schug, K.A., Pretreatment techniques for produced water with subsequent forward osmosis remediation. Water, 2019, 11, 1437-1449.

[107] Patil, O., Sayyad, S.U., Forward osmosis application in treatment of wastewater. International Journal of Engineering Trends and Technology, 2016, 37(4), 233-239.

[108] Jafarinejad, S., Park, H., Mayton, H., Walker, S.L. and Jiang, S.C., Concentrating ammonium in wastewater by forward osmosis using a surface modified nanofiltration membrane. Environmental Science: Water Research and Technology, 2019, 5, 246-255.

[109] Xu, S., Lin, P., An, X.,, Hu, Y., Wang, Z., Zhong, L., Niu, Q., High-performance forward osmosis membranes used for treating high-salinity oil-bearing wastewater. Ind. Eng. Chem. Res., 2017, 56, 12385-12394.

[110] Zhu, S., Li, M., El-Din, M.G., Forward osmosis as an approach to manage oil sands tailings water and on-site basal depressurization water. J. Hazard. Mater., 2017, 327, 18-27.

[111] Volpin, F., Chekli, L., Phuntsho, S., Cho, J., Ghafour, N., Vrouwenvelder, J.S., Kyong Shon, H., Simultaneous phosphorous and nitrogen recovery from source-separated urine: A novel application for fertiliser drawn forward osmosis. Chemosphere, 2018, 203, 482-489.

[112] Gwak, G., Kim, D.I., Hong, S., New industrial application of forward osmosis (FO): Precious metal recovery from printed circuit board (PCB) plant wastewater. J. Membr. Sci., 2018, 552, 234-242.

[113] Buckwalter, P., Forward osmosis for wastewater treatment and energy recovery: a techno-economic analysis, Humboldt State University, USA Master of Science in Environmental Systems: Energy Technology and Policy, 2018, 56.

[114] Mondal, D., Nataraj, S.K., Reddy, A.V.R., Ghara, K.K., Maiti, P., Upadhyay, S.C., Ghosh, P.K., Four-fold concentration of sucrose in sugarcane juice through energy effcient forward osmosis using sea bittern as draw solution. RSC ADV., 2015, 5, 17872-17878.

[115] Gawande, N., Mungray, A.A. Superabsorbent polymer (SAP) hydrogels for protein enrichment. Sep. Purif. Technol., 2015, 150, 86-94.

[116] Zaviska, F., and Zou, L., Using modelling approach to validate a bench scale forward osmosis pre-treatment process for desalination. Desalination, 2014, 350: 1-13.

[117] Bamaga, O. A., Yokochi, A., Zabara, B., and Babaqi, A.S. Hybrid FO/RO desalination system: preliminary assessment of osmotic energy recovery and designs of new fo membrane module configurations. Desalination, 2011, 268: 163-169.

[118] Zhang, X.H., Li, Q.G., Wang, J., Li, J., Zhao, C.W., Hou, D.Y., Effects of feed solution pH and draw solution concentration on the performance of phenolic compounds removal in forward osmosis process. J. Environ. Chem. Eng., 2017, 5, 2508-2514.

[119] Altaee, A., Zaragoza, G., and van Tonningen, H.R. Comparison between Forward Osmosis-Reverse Osmosis and Reverse Osmosis processes for seawater desalination. Desalination, 2014, 336, 50-57.

[120] Yangali-Quintanilla, V., Li, Z., Valladares, R., Li, Q., and Amy, G. Indirect desalination of red sea water with forward osmosis and low pressure reverse osmosis for water reuse. Desalination, 2011, 280: 160-166.

[121] Choi, B.G., Zhan, M., Shin, K., Lee, S., Hong, S., Pilot-scale evaluation of FO-RO osmotic dilution process for treating wastewater from coal-fired power plant integrated with seawater desalination. J.Membr. Sci. 2017, 540, 78-87.

[122] Zhao, S., Zou, L., and Mulcahy, D., Brackish Water desalination by a hybrid forward osmosis–nanofiltration system using divalent draw solute. Desalination, 2012, 284: 175-181.

[123] Tan, C. H., and Ng, H. Y., A novel hybrid forward osmosis-nanofiltration (FO-NF) process for seawater desalination: draw solution selection and system configuration. Desalination and Water Treatment. 2010, 3: 356-361.

[124] Giagnorio M., Ricceri, F., Tagliabue, M., Zaninetta, L., and Tiraferri, A., Hybrid forward osmosis–nanofiltration for wastewater reuse: system design. Membranes, 2019, 9(5), 61-74.

[125] Zou, S., He, Z., Electrodialysis recovery of reverse-fluxed fertilizer draw solute during forward osmosis water treatment. Chem. Eng. J., 2017, 330, 550-558.

[126] Bitaw, T.N., Dae, K.P., Yang, R., Optimization on a new hybrid forward osmosis-electrodialysis-reverse osmosis seawater desalination process. Desalination, 2016, 398, 15, 265-281.

[127] Zhang, Y., Pinoy, L., Meesschaert, B., der Bruggen, B.V., A natural driven membrane process for brackish and wastewater treatment: photovoltaic powered ED and FO hybrid system. Environ Sci Technol., 2013, 47(18), 10548-10555.

[128] Kwon, K., Han, J., Park, B.H., Y., Shin, Kim, D., Brine recovery using reverse electrodialysis in membrane-based desalination processes. Desalination, 2015, 362, 1-10.

[129] Zhou, Y., Huang, M., Deng, Q., Cai, T., Combination and performance of forward osmosis and membrane distillation (FO-MD) for treatment of high salinity landfill leachate. Desalination, 2017, *420*, 99-105.

[130] Salih, H.H., Dastgheib, S.A., Treatment of a hypersaline brine, extracted from a potential CO_2 sequestration site, and an industrial wastewater by membrane distillation and forward osmosis. Chem. Eng. J., 2017, 325, 415-423.

[131] Peñate, B., and García-Rodríguez, L., Current trends and future prospects in the design of seawater reverse osmosis desalination technology. Desalination, 2012, 284, 1-8.

[132] Liu, C., Rainwater, K., and Song, L., Energy analysis and efficiency assessment of reverse osmosis desalination process. Desalination, 276, 352-358.

[133] Blandin, G., Verliefde, A.R.D., Comas, J., Rodriguez-Roda, I., Le-Clech, P., Efficiently combining water reuse and desalination through forward osmosis reverse osmosis (FO-RO) hybrids: A critical review. 2016, Membranes 6 (3).

[134] Blandin, G., Verliefde, A.R.D., Tang, C.Y., Le-Clech, P., Opportunities to reach economic sustainability in forward osmosis-reverse osmosis hybrids for seawater desalination. Desalination, 2015, 363, 26-36.

[135] Kim, J.E., Phuntsho, S., Chekli, L., Choi, J.Y., Shon, H.K., 2018. Environmental and economic assessment of hybrid FO-RO/NF system with selected inorganic draw solutes for the treatment of mine impaired water. Desalination 429, 96-104.

[136] Pal, P., Chakrabortty, S., Nayak, J., Senapati, S., A flux-enhancing forward osmosis-nanofiltration integrated treatment system for the tannery wastewater reclamation. Environ. Sci. Pollut. Res., 2017, 24, 15768-15780.

[137] Tan, C. H., Lefebvre, Zhang, O., Ng, H.Y., Ong, S.Y., membrane technology and environmental a pplications: membrane processes for desalination: overview. 2012, ASCE, Chapter 10 p. 27.

[138] Qin, M., Hynes, E.A., Abu-Reesh, I.M., He, Z., Ammonium removal from synthetic wastewater promoted by current generation and water flux in an osmotic microbial fuel cell. J. Clean. Prod., 2017, 149, 856-862.

[139] Zhu, W., Wang, X., She, Q., Li, X., Ren, Y., Osmotic membrane bioreactors assisted with microfiltration membrane for salinity control (MF-OMBR) operating at high sludge concentrations: performance and implications. Chem. Eng. J. 2018, 337, 576-583.

[140] Zhang, F., Brastad, K.S., He, Z., Integrating forward osmosis into microbial fuel cells for wastewater treatment, water extraction and bioelectricity generation, Environ. Sci. and Tech., 2011, 45, 6690-6696.

[141] Li, W.W., Yu, H.Q., He, Z., Towards sustainable wastewater treatment by

using microbial fuel cells-centered technologies. Energ. Environ. Sci., 2014, 7, 911-924.

[142] He, Z., One more function for microbial fuel cells in treating wastewater: Producing high-quality water. Chemik, 2012, 66, 7-10.

[143] Ge, Z., He, Z., Effects of draw solutions and membrane conditions on electricity generation and water flux in osmotic microbial fuel cells. Bioresour. Technol., 2012, 109, 70-76.

[144] Ge, Z., Ping, Q.Y., Xiao, L., He, Z., Reducing e_uent discharge and recovering bioenergy in an osmotic microbial fuel cell treating domestic wastewater. Desalination, 2013, 312, 52-59.

[145] Liu, J., Wang, X., Wang, Z., Lu, Y., Li, X., Ren, Y., Integrating microbial fuel cells with anaerobic acidification and forward osmosis membrane for enhancing bio-electricity and water recovery from low-strength wastewater. Water Research, 2017, 110, 74-87.

[146] Lu, Y.B., Qin, M.H., Yuan, H.Y., Reesh, I.A., He, Z., When bioelectrochemical systems meet forward osmosis: Accomplishing wastewater treatment and reuse through synergy. Water, 2014, 7, 38-50.

[147] Qiu, G., Ting, Y.-P., Direct phosphorus recovery from municipal wastewater via osmotic membrane bioreactor (OMBR) for wastewater treatment. Bioresour. Technol., 2014, 170, 221-229.

[148] Zhu, W., Wang, X., She, Q., Li, X., Ren, Y., Osmotic membrane bioreactors assisted with microfiltration membrane for salinity control (MF-OMBR) operating at high sludge concentrations: Performance and implications. Chem. Eng. J., 2018, 337, 576-583.

[149] Luo, W., Hai, F.I., Price, W.E., Guo, W., Ngo, H.H., Yamamoto, K., Nghiem, L.D., Phosphorus and water recovery by a novel osmotic membrane bioreactor–reverse osmosis system. Bioresour. Technol., 2016, 200, 297-304.

[150] Aftab, B., Khan, S.J., Maqbool, T., Hankins, N.P., Heavy metals removal by osmotic membrane bioreactor (OMBR) and their effect on sludge properties. Desalination, 2017, 403, 117-127.

Chapter 4

Integration of Forward Osmosis in Municipal Wastewater Treatment Applications

Stavroula Kappa and Simos Malamis

Abstract

In recent years, the research community has made constant efforts to develop new technologies for the recovery and valorization of water, nutrient and energy content of municipal wastewater. However, the recovery process is significantly limited due to the low-strength of sewage. Over the last 10 years, the Forward Osmosis (FO) process, has gained interest as a low-cost process with low membrane fouling propensity, which can convert municipal wastewater into a concentrated low-volume effluent, characterized by high organic and nutrient concentration. This chapter presents the main configurations that have been implemented for the concentration of municipal wastewater using FO, including their performance in terms of contaminant removal and water/reverse salt flux (J_w/J_s). Furthermore, the draw solutions and respective concentrations that have been used in FO for the treatment of sewage are reported, while at the same time the positive and negative characteristics of each application are evaluated. Finally, in the last section of this chapter, the spontaneous FO followed by anaerobic process is integrated in a municipal wastewater treatment plant (WWTP) and compared with a conventional one. The comparison is done, in terms of the mass balance of the chemical oxygen demand (COD) and in terms of the energy efficiency.

Keywords: forward osmosis, municipal wastewater, configurations, draw solution, COD mass balance

1. Introduction

Water scarcity is one of the most serious threats which our planet faces [1]. Globally, water demand is predicted to increase by 35% more than sustainable supply by 2040/50, if the linear water management model continues to be implemented [2]. The European Union (EU) encourages the implementation of a circular economy model, through its strategy called "Closing the loop—a EU action plan for the Circular Economy" in 2015 and European citizens must seize the opportunity to close the loop of water, resource and energy management [3]. Among various types of water, seawater and wastewater are two alternative sources, which are readily available, especially in coastal, arid areas [4]. Both need to be treated before they can be rendered suitable for use. Membrane processes such as microfiltration (MF), ultrafiltration (UF), nanofiltration (NF) and reverse osmosis (RO) are particularly effective in the purification of non-conventional water sources and count many applications in both the wastewater and the desalination sector [5]. In particular, the RO, holds a prominent position in water desalination, compared to traditional thermal desalination processes [6]. As high energy consumption is required to overcome the osmotic potential, reverse osmosis is not applied in many water-stressed areas [5].

Forward osmosis is one of the most attractive membrane-based processes that requires two solutions of different osmotic concentrations (high and low), separated by a semi-permeable membrane to be realized [7]. Water molecules are spontaneously diffused from the low osmotic potential solution (feed side) to the high osmotic solution (draw solution or DS), to equalize the concentration difference, while the semi-permeable membrane acts as a barrier that rejects the salts and contaminants [8]. The natural osmotic pressure of FO makes it stand out from conventional RO, by offering high water recovery, reduced membrane fouling potential, greater effectiveness, low cost, and reduced energy demand [8, 9]. All these positive aspects have led to a notably high trend of publications on FO applications in various water sources, such as seawater and wastewater, with more than 97.5% of publications since 2009 [10]. Among them, several researchers investigate the feasibility of integrating the FO process in a novel sewage treatment system based on the circular economy concept, as the main goal is to valorize the chemical energy, water, and nutrients of sewage. This innovative application of FO and its combination with appropriate downstream technologies is really promising. As the results show, the wastewater is converted into a small volume liquid, characterized by a high concentration of organic matter, as it can be concentrated up to 8–10 times, while the recovery of phosphorus can reach up to 90%, replacing the need for chemical fertilizers [11, 12]. However, there are many challenges that need to be overcome for this application, the most important of which is the selection of the most appropriate DS, which despite the significant efforts has not been found to date [13–15].

This chapter presents the main configurations that have been implemented to concentrate municipal wastewater using FO, including their performance in terms of contaminant removal and J_w/J_s. The draw solutions and their concentrations that have been used in the FO process for the treatment of sewage are reviewed, while at the same time the positive and negative characteristics of each application are evaluated. Finally, in the last section of this chapter, the spontaneous FO followed by an anaerobic process is integrated into a municipal wastewater treatment plant and compared with a conventional activated sludge process (CAS), in terms of COD and corresponding energy efficiency, emphasizing the key impact of the FO in the latter process.

2. Forward osmosis configurations and performance in municipal wastewater management

The main benefit of the FO process in municipal wastewater treatment is that it converts sewage from low-strength liquid to a concentrated bulk, which consists of high a concentration of organic matter and nutrients [16, 17]. According to Korenak et al. [18], the FO process is characterized by high membrane fouling reversibility, while it can significantly minimize space requirements in a municipal WWTP. Considering all the above, three basic configurations have emerged for the integration of the FO process in the municipal WWTPs, which are illustrated in **Figure 1**.

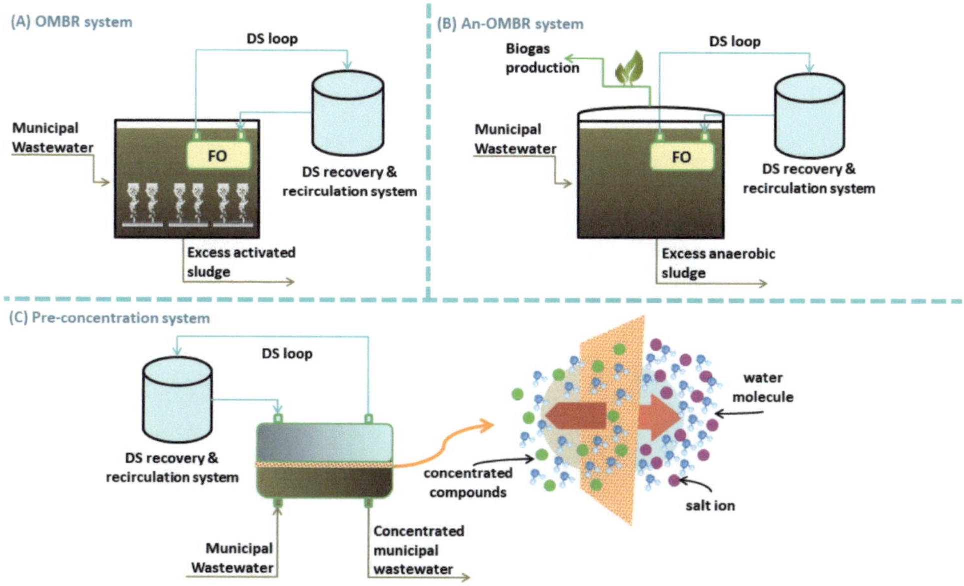

Figure 1.
Configurations for the integration of the FO process in the municipal WWTPs.

2.1 Osmotic membrane bioreactor (OMBR)

In 2008, an innovative system was introduced, in which FO membranes were submerged into a typical membrane bioreactor (MBR) module; this system was called OMBR (**Figure 1(A)**) [19]. The replacement of UF or MF membranes in the conventional system by FO membranes resulted in better performance in terms of contaminants' rejection (79.7–100% of COD, **Table 1**). In addition, the absence of hydraulic pressure contributed to lower fouling tendency and probably lower energy requirements. Despite the benefits of OMBR over traditional systems, two major challenges are still under investigation; low J_w rate and salinity accumulation [19, 31]. The findings confirm that the decline in J_w was greatly affected by the salt accumulation, even with the implementation of improved membrane materials, such as thin-film composite (TFC), achieving an average rate equal to 3.9 ± 0.5 L m^{-2} h^{-1} [32]. In addition, the microbial community of the reactor can either be partly or fully inhibited, due to the gradual building-up of salts, which occurs due to the J_s [31, 32].

2.2 Anaerobic OMBR (An-OMBR)

The combination of MBR technology with the anaerobic process has been extensively investigated in the last 10 years, due to the environmental benefits of both [33]. However, the low-strength nature of sewage is a major obstacle to the effective application of the anaerobic process in municipal WWTPs; containing a high amount of water with low organic and nutrients concentration. Due to the methane's solubility in water (22.7 mg L^{-1}, at room temperature), a large part of the produced gas escapes with the treated effluent of the anaerobic process (ranges between 20 and 60%) [34]. Due to the aforementioned barriers, it is difficult to implement anaerobic processes for municipal wastewater treatment particularly in areas, where the sewage temperature drops below 15°C, during the winter period. The incorporation of FO, either as a pre-treatment step or submerged into the MBR system, significantly enhances the resource recovery potential in the anaerobic

Membrane Type	Feed	Draw	Removal Efficiency (%)					J_w ($L\ m^{-2}\ h^{-1}$)/J_s ($g\ m^{-2}\ h^{-1}$)	Ref.
			TOC	COD	TP	NH_4-N	TN		
CTA (HTI, USA)	Municipal wastewater	Seawater brine	—	90	97.9	99	—	N.A.	[12]
CTA (HTI, USA)	Synthetic municipal wastewater	1 M NaCl	98 ± 1	—	—	99 ± 1	—	8–5/NA.	[20]
TFC (HTI, USA)	Synthetic municipal wastewater	1 M NaCl	96 ± 1	—	—	99 ± 1	—	3/N/A	[20]
CTA (HTI, USA)	Municipal wastewater	42 $g\ L^{-1}$ NaCl	—	99	99	—	82	4–4.5/ N.A.	[21]
CTA-ES (HTI, USA)	Sunthetic municipal wastewater	48.4 $g\ L^{-1}$ $MgCl_2$	98	—	—	98	—	3.7–3.3/N.A	[22]
CTA (HTI, USA)	Raw municipal wastewater	Synthetic seawater	79.7 ± 9	—	92 ± 3.3	88.1 ± 5.5	—	4.86–3.24/ N.A	[23]
CTA-ES (HTI, USA)	Sunthetic municipal wastewater	48.4 $g\ L^{-1}$ $MgCl_2$ & 49 $g\ L^{-1}$ NaCl	98	—	98 (PO_4^{3-}-P)	98	—	6.64–8.95 (NaCl); 6.46 ($MgCl_2$)/ NA.	[24]
TFC (HTI, USA)	Sunthetic municipal wastewater	1 M NaCl	100	—	95.6 (PO_4^{3-}-P)	43		14/ N.A.	[25]
CTA (HTI, USA)	Raw sewage	0.5 M NaCl	95	—	95	95	90	1.7/ NA.	[26]
CTA & TFC (HTI, USA), Biomimetic (Aquaporin, Denmark)	Synthetic wastewater	0.5 M NaCl.	>95	—	—	90–100	60–80	15.6–15 (TFC); 5.5 (CTA)/0.085 (Aquaporin); 82.7 (TFC); 5.5 (CTA) mmol $m^{-2}\ h^{-1}$	[27]
CTA-ES (HTI, USA)	Synthetic wastewater	1 M $MgCl_2$	—	—	99	99	—	5.45/ N.A.	[28]
TFC (Aquaporin A/S, Asia)	Synthetic wastewater	1.2 M NaCl with SDBS	100	—	>90	85	50–80	10/ N.A	[29]
TFC (Aquaporin A/S, Asia)	Synthetic wastewater	1.2 M NaCl without SDBS	98	—	>90	>90	60–100	<10/ N.A.	[29]
CTA-ES (HTI, USA)	Synthetic municipal wastewater	70 $g\ L^{-1}$ NaCl	—	99	99 (PO_4^{3-}-P)	92	—	10.42–6.4/9.9–34.99	[30]

Table 1.
Osmotic Membrane Bioreactor applications to treat municipal wastewater.

process. Compared to OMBR, An-OMBR (**Figure 1(B)**) is characterized by significantly lower energy requirements, due to the replacement of energy-demanding aeration, while biogas production contributes to the coverage of specific energy needs. According to Zhang et al. [35], due to the 2–3 times smaller pore size of the FO membranes over conventional UF or MF membranes, the dissolved methane content in the An-OMBR treated effluent was eliminated, even as a function of different operating parameters. Regarding the yield of methane, Zhang et al. [35] and Gu et al. [36] observed particularly satisfactory production that reached 0.256 L CH_4 g^{-1} COD and 0.25–0.3 L CH_4 g^{-1} COD at mesophilic conditions, respectively. In addition, anaerobic biomass showed high resistance to increasing salt concentrations and was not affected even when the concentration was equivalent to 200 mM sodium chloride (NaCl) [36]. As shown in **Table 2** the FO membranes achieve high rejection of contaminants; specifically the Total Organic Carbon (TOC)/COD and PO_4-P removal was 95% and 73%, respectively. However, due to the lack of ammonia removal, its accumulation has been observed in the reactor, but not in concentrations that can lead to the interruption of the anaerobic process [36]. In recent years, an alternative configuration has been proposed, which includes the addition of MF membranes both to OMBR and An-OMBR systems, the so-called Microfiltration- Osmotic Membrane Bioreactor (MF-OMBR). The main goal of this hybrid system is 1) to balance the salts concentration in the reactor so as to prevent an inhibition event and 2) to apply resource recovery methods to its nutrient-rich

Membrane Type	Feed	Draw	Removal Efficiency (%)					Jw (L m^{-2} h^{-1})/Js (g m^{-2} h^{-1})	Ref.
			TOC	COD	TP	NH_4-N	TN		
CTA (HTI, USA)	Synthetic municipal wastewater	0.5 M NaCl	—	95	99	70–80	—	10–3/ N.A.	[36]
CTA (HTI, USA)	Synthetic municipal wastewater	0.5 M NaCl	—	96	100	62	—	9.5–3.5/ N.A.	[37]
TFC (Aquaporin Denmark)	Synthetic municipal wastewater	116.6 g L^{-1} $MgSO_4$	—	>95	95 (PO_4^{3-}-P)	>95	—	0.78–0.26/ NA.	[38]
CTA (HTI, USA)	Synthetic domestic wastewater	0.5 M NaCl	>96	—	—	—	—	6–3.4/ N.A.	[39]
CTA-ES (HTI, USA)	Municipal sewage	0.5 M, 1 M and 1.5 M NaCl	—	96	100	88	89	6 (0.5 M); 10 (1 M); 13 (1.5 M)/ 4.26 (0.5 M); 7.65 (1 M); 11.84 (1.5 M)	[40]
CTA-NW (HTI, USA)	Synthetic municipal wastewater	2 M $C_4H_6MgO_4$	—	96	73 (PO_4^{3-}P)	51.4	—	3.5–1.09/ 2.5–1.6	[35]
TFC (HTI, USA)	Synthetic municipal wastewater	0.5 NaCl	—	>95	99 (PO_4^{3-}P)	0	—	12 ± 0.7–2 ± 0.2/ N.A.	[41]
CTA (HTI, USA)	Synthetic sewage	0.5 NaCl	—	>93	99	28–45	—	8.7 ± 0.3–4.0 ± 0.2/ N.A.	[42]

Table 2.
Anaerobic Osmotic Membrane Bioreactor applications to treat municipal wastewater.

treated effluent. Nonetheless, according to Wang et al. [39], the FO membranes achieved much lower ammonia rejection rates (39–50%) compared to an An-OMBR system (62.7–81.2%), while the addition of another membrane significantly raises both the maintenance and the investment cost of the entire system [39, 43].

2.3 Pre-concentration with FO

Alternatively, the FO unit can be applied as a pre-condensation step in municipal WWTPs (**Figure 1(C)**), achieving a similar goal to the previously analyzed configuration, as it can be combined by suitable downstream processes for resources and energy utilization. As reported by Ansari et al. [34], the submerged FO configuration is significantly disadvantaged compared to the separate one, as the former gets in contact with the dense activated sludge, while the latter with the diluted primary treated effluent. In contrast, a recent study that examined both approaches in parallel, direct osmosis showed a significant decline in J_w performance compared to OMBR system [7]. On the other hand, a prolonged biodegradation study (approximately 7 months) of both cellulose triacetate (CTA) and TFC membranes demonstrated that the long-term exposure to activated sludge significantly affects their performance, in terms of water permeability and J_s [44]. Sun et al. [7] found that the direct FO module is characterized by reversible membrane fouling over the submerged OMBR membrane, mainly due to the lower abundance in the microbial load of the feed solution. In terms of performance, as shown in **Table 3**, this FO configuration achieves the retention of organic load by a percentage ranging from 71.9 to 100%. At this point, it should be noted that based on the current literature most studies refer to FO as either a separate or integrated system of an Anaerobic Membrane Bioreactor (An-MBR), while alternative anaerobic treatment systems are not frequently investigated.

3. Draw solutions

In contrast to other osmotic, membrane-based technologies, the application of high osmotic potential is the driving force in the spontaneous FO process [52]. Therefore, the selection process of the most effective solution acts as a cornerstone of the FO and plays a crucial role on its performance as well as on downstream processes [15]. In an ideal physicochemical context, the parameters listed in **Table 4** must be met to classify a solution as appropriate [52–54].

In recent years, significant efforts have been made by researchers to combine the above parameters and develop an ideal DS, which will be compatible with the application of FO in the municipal wastewater treatment sector [15, 55, 56]. Alternative systems have been developed; different configurations have been applied to integrate the FO in several stages of a municipal WWTP; as pre-treatment, secondary and post-treatment steps for nutrient recovery. Obviously, the treatment level and the quality-target of the recovered product must be considered in the DS selection process [57]. First on the list and most commonly used as DS is NaCl, even in high concentrations up to 4 M, due to its high aqueous solubility, small molecular size, high availability, and relatively low cost [58]. As shown in **Table 5**, the 0.5 M concentration is most frequently applied, as it simulates the osmotic pressure of seawater [53]. The ultimate goal is to adopt a circular solution, by applying an abundant water source without any economic burden or a process' by-product, such as the RO brine as DS (**Table 6**) [58, 65]. High rejection rates of TOC/COD and PO_4-P have been reported using NaCl as DS in OMBR systems, equal to 100% and 95.6%, respectively, although the same is not achieved for ammonium nitrogen

Membrane Type	Feed	Draw	Removal Efficiency (%)					Jw ($L\ m^{-2}\ h^{-1}$)/Js ($g\ m^{-2}\ h^{-1}$)	Ref.
			TOC	COD	TP	NH_4-N	TN		
CTA (HTI, USA)	Raw Sewage	3.5% NaCl	—	71.9 (AL-FS); 69.3 (AL-DS)	—	—	—	5.2 (AL-DS) and 5.4 (AL-FS)/ N.A.	[45]
CTA (HTI, USA)	Municipal wastewater	0.5 M NaCl	—	99.8 ± 0,6	99.7 ± 5	48.1 ± 10.5	67.8 ± 7.3	7.7–6.5/ 5	[17]
TFC, (Aquaporin, Denmark)	Domestic sewage	1, 1.5 & 2 M $MgCl_2{\cdot}6H_2O$	—	76–80	75	66	—	5.11 (1 M); 6.66 (1.5 M); 6.92 (2 M)/ N.A.	[16]
CTA (HTI, USA)	Sewage	0.2–4 M NaCl	—	96.5	95.4	93.3	96.5	10 (0.5 M) - 25 (4 M)/ N. A.	[46]
CTA (HTI, USA)	Municipal wastewater	Synthetic seawater	79.9 ± 6.7	—	93.3 ± 3.3	85.4 ± 5.6	—	5.37–2.5/ N.A.	[23]
TFC (Toray, Japan)	Municipal wastewater	11.7 g L^{-1} NaCl	—	85.5	90	no rejection (AL-FS); 50 (AL-DS)	—	5.1 ± 1 / 4.8 ± 8.6	[47]
Aquaporin TFC (Sterlitech Corporation, WA)	Synthetic wastewater	Synthetic seawater with *C. vulgaris*	—	100 ± 0.6	99.5 ± 0.5	46.1 ± 3.4	—	11.59 ± 0.49	[48]
Aquaporin TFC (Sterlitech Corporation, WA)	Synthetic wastewater	Synthetic seawater without C. vulgaris	—	99.5 ± 0.7	99.1 ± 0.1	46.2 ± 2.6	—	12.02 ± 0.35	[48]
TFC (HTI, USA)	Municipal wastewater &prefiltered	Synthetic seawater brine	—	2.38–2.67 CF	3.3–3.5 CF	1.31–1.75 CF	1.58–1.94 CF	19.90 (raw sewage); 18.15 (filtered)/ N.A.	[49]
CTA (HTI, USA)	Synthetic municipal wastewater	4% NaCl	—	99	—	67–68	56–59	N.A.	[50]
CTA-ES (HTI, USA)	Synthetic municipal wastewater	0.25 M $(NH4)_2SO_4$ (SOA), KH_2PO_4 (MKP),$(NH_4)H_2PO_4$ (MAP)	—	98	99	89,56 ± 1.52 (SOA); 98.96 ± 0.33 (MPK); 89.67 ± 1.27 (MAP)	—	2.58/ 0.57 (SOA); 2.11/ 1.17 (MPK); 1.97/ 0.11 (MAP)	[51]

Table 3.
Pre-concentration of municipal wastewater using FO.

Parameter	Impact
Osmotic Pressure	DS with higher osmotic pressure than the feed generates higher J_w
Water Solubility	Soluble compound produces higher osmotic concentration and therefore retains J_w and water recovery at higher percentages
Concentration	Higher concentrations contribute to higher J_w rates, but particularly high concentrations inhibit flux's increase, as mass transfer phenomena occur, such as concentration polarization
Diffusion, viscosity and molecular weight	Small molecules are distinguished by high aqueous solubility and high osmotic pressures, but in comparison with the large ones they are characterized by a higher diffusion coefficient and J_s
Toxicity and degradation	Low reverse salt diffusion to minimize the risk of toxicity and contamination of downstream systems and the recovered product. Low rate of degradability, unless it is beneficial
Cost-effective	Easily re-concentration at competitive cost
Availability	Available in large quantities with low price and easy handling

Table 4.
Main parameters that characterized the ideal DS.

(NH_4-N), which in most studies ranges between 43 and 90% [25, 59]. Nevertheless, the biggest challenge in OMBR systems using NaCl as DS is the accumulation of salts in the concentrated stream of mixed liquor and the subsequent negative effect on bacterial growth, due to reverse sodium leakage [12]. Relevant mitigation measures of the above obstacles have been proposed, such as the reduction of sludge retention time (SRT), but also the application of hybrid solutions, such as MF and UF membranes downstream for the parallel recovery of phosphorus [32].

Similar results are demonstrated in bench and pilot scale FO systems for the pre-concentration of municipal wastewater using NaCl. The bidirectional diffusion of monovalent ammonium ions from the feed to the sodium cations of DS remains a major drawback [17]. In a recent study, Yang et al. [49] demonstrated the effect of the pH parameter on low NH_4-N rejection rates and suggested a functional range of less than 8 for optimized performance. More specifically, at elevated pH as the main form of ammonium nitrogen is ammonia, diffusion becomes independent of the reverse sodium leakage [49]. Alternatively, the application of divalent molecular compounds as DS (**Tables** 7 and **8**), such as magnesium chloride ($MgCl_2$) and magnesium sulfate ($MgSO_4$), which are characterized by lower reverse salt transport than NaCl, is suggested in many investigations [16]. Another superiority of inorganic solutions containing Mg ions is their combination with MF-OMBR hybrid systems and the utilization of the reverse Mg flux in the mixed liquor to nutrients' recovery, after proper pH adjustment. Although, a comparative study demonstrated that Mg transport leads to the formation of both organic and inorganic fouling in the active and support layer of the TFC membrane, correspondingly, causing a dramatic reduction in membrane flux [56]. As shown in **Table 8**, a highly charged compound, ethylenediamine tetraacetic acid disodium salt (EDTA 2Na) was applied as DS to remove the water from the activated sludge in a hybrid Forward Osmosis – Nanofiltration (FO-NF) system; the NF module was used for the recovery of DS. Water flux dropped rapidly after 8 operating hours (8.45 to 4.22 L m^{-2} h^{-1}), mainly due to the reduction of the osmotic driving force and the formation of a cake layer on the membrane surface. It is worth noting that the reverse salt flux was equal to 0.2 g m^{-2} h^{-1}, while suspended solids were concentrated from 8 g L^{-1} to 32 g L^{-1} [75].

Draw	Feed	Configuration/ Membrane Type	Findings	Ref.
0.5, 1 & 1.5 M NaCl	Municipal Wastewater	An-OMBR/CTA-ES (HTI,USA)	• Salinity accumulation, 1 M NaCl was advantageous	[40]
0.5 M NaCl	Municipal Wastewater	Direct FO/ Spiral Wound CTA (HTI, USA)	• High contaminant rejection rates, except for ammonia	[17]
0.5 M NaCl	Synthetic Wastewater	OMBR/ CTA (HTI, USA)	• Accumulation of nutrients and salts in the OMBR	[59]
3.5 g L^{-1} NaCl	Domestic wastewater	Direct FO (coupled with MD)/ CTA (HTI, USA)	• Fouling of organic substances such as proteins and polysaccharides, resulting in reduced J_w	[60]
0.5 M NaCl	Synthetic municipal wastewater	OMBR/ Biomimetic (AQUAPORIN Asia, Singapore) TFC & CTA (HTI,USA)	• Aquaporin FO membrane showed better performance than CTA & TFC, in terms of salinity accumulation	[27]
3 M NaCl	Synthetic wastewater	Post FO/ TFC-ES (HTI, USA)	• J_w decline due to the dense adsorption layer and the gel layer formed by the deposition of carbohydrates and proteins	[61]
30 g L^{-1} NaCl	Algae effluent	Algae - hybrid FO-RO system/ TFC Porifera Inc. (California, USA)	• The FO rejected organic, multivalent cations and anions, providing an effective pretreatment for the RO system	[62]
0.2, 0.5, 1, 1.5, 2, 3 & 4 M NaCl	Municipal wastewater	Direct FO/ CTA-ES (HTI,USA)	• Disproportionate concentration between contaminants and water	[46]
			• Low fouling propensity, with layers formed by humic acid, protein, and polysaccharide	[46]
1 M NaCl	Synthetic municipal wastewater	OMBR/ TFC (HTI, USA)	• NH_4-N rejection rate was low	[25]
			• The water flux decreased with increasing salinity in the mixed liquor	[25]
70 g L^{-1} NaCl	Raw anaerobic centrate	Direct FO-RO/CTA (HTI,USA)	• FO/RO system affected by centrate replenishment and concentration; it decreased with increased replenishment and concentrations	[63]
0.7 g L^{-1} NaCl & 0.7 g L^{-1} Na_2SO_4	Synthetic wastewater	AnFOMBR/ CTA (HTI, USA)	• NaCl recorded better methane ratio in the biogas produced	[64]
53 g L^{-1} NaCl and Industrial effluent (consisted of SO_4^{2-} and NH_4-N)	Synthetic wastewater	OMBR/ CTA-NW	• Industrial wastewater had higher J_w and less membrane fouling compared to NaCl, but, a higher J_s was observed in the former	[65]
0.5 M NaCl & NaCl with 0.01 M ($(NH_4)_2HPO_4$)	MBR permeate	FO-RO/TFC, FTSH2O (Sterlitech Company, USA)	• The enhanced DS with DAP recorded higher J_w than NaCl Higher rejection of TP than NH_4-N	[66]

Table 5.
Sodium chloride as DS in FO treating municipal wastewater.

Draw	Feed	Configuration/ Membrane Type	Findings	Ref.
Brine	Primary municipal effluent & raw municipal wastewater	Direct FO/ homemade TFC	• J_w depends on temperature, high viscosity at low temperatures increase ICP effect • Reversible fouling after physical cleansing	[49]
Synthetic seawater	Synthetic feed and MBR permeate	Direct FO/ TFC, CTA-ES and CTA-NW, (HTI,USA)	• Concentration of nutrients successfully performed • pH plays a key role in ammonia rejection by the FO membranes (close to 7)	[67]
Brine and industrial effluent (mainly consisted of SO_4^{2-} and NH_4-N)	Anaerobically digested sludge centrate	Direct FO/ CTA-NW (HTI,USA) and Aquaporin (AQUAPORIN A/S, Denmark)	• 2 industrial effluents successfully implemented as DS in the FO process • Increased NH_4-N concentration in concentrated sludge due to application of DS rich in ammonium sulfate	[68]
Seawater	Anaerobically digested sludge centrate	Direct FO / TFC, Porifera, Inc. (California, USA)	• Extensive membrane fouling due to nutrient precipitation in both feed solution and membrane • Filtration time plays an essential role in process performance	[69]
Synthetic seawater & brine	Wastewater after a hydrolytic anaerobic reactor	Direct FO / TFC Porifera Inc. (California, USA)	• With seawater & brine the condensation factor can reach over 10 • Inevitable biodegradation of VFAs in this environment	[70]
Synthetic seawater with algae strain	Synthetic municipal wastewater	Direct FO / Aquaporin TFC (Sterlitech Corporation, WA)	• Low ammonium rejection, low removal of ammonia after the application of algae at about 35%.	[48]
Seawater (0.599 M, 0.428 M & 0.770 M NaCl solutions)	Anaerobically digested sludge centrate	Direct FO / Aquaporin (Aquaporin A/S, Denmark)	• Better performance at pH <9 and application of DS with low reverse salt	[63]
Brine	Raw municipal wastewater	MF-FOMBR/ CTA (HTI, USA)	• 90% recovery of phosphorus using the MF system • Accumulation of salts in the bioreactor is still a challenge	[12]

Table 6.
Seawater, Brine, and industrial effluents as DS in FO treating municipal wastewater.

To enhance the valorization of the resources contained in municipal wastewater, through the application of the anaerobic process several organic and ionic organic draw solutions have been investigated [13, 14, 74]. Bowden et al. [14] compared 10 different ionic organic compounds as DS and slightly altered the selection methodology proposed by Achilli et al. [15], introducing the parameter of biodegradability

Draw	Feed	Configuration/ Membrane Type	Findings	Ref.
0.5, 1, 1.5 & 2 M $MgCl_2$	Synthetic Secondary effluent	Post FO / CTA-NW and CTA-ES (HTI,USA)	• Cl diffusion was higher compared to Mg ions. 95% rejection of nutrients using 2 M $MgCl_2$ • Higher diffusion of Cl (about 3 times) by applying CTA-ES membrane	[71]
3000:1; 1500:1: 1000:1 $MgCl_2$ &Triton X-144	Synthetic domestic wastewater	SMB-OSMBR/ CTA-ES (HTI, USA)	• 1.5 mM $MgCl_2$ and 1.5 mM Triton X-114 was the best solution ratio in terms of performance • Biomass growth media favored the achievement of stable J_w and low membrane fouling	[72]
1 M $MgCl_2$	Synthetic municipal wastewater	OMBR-MD/ CTA - ES (HTI, USA)	• Successful rejection of PO_4-P and NH_4-N & recovery in the form of struvite • Recovery of the DS with the MD system with a small drop in the J_w	[28]
1 M $MgCl_2$	Raw anaerobic centrate	Direct FO-MD/ CTA (HTI,USA)	• High rejection of nutrients, reversible membrane fouling, potential recovery of struvite with the application of FO-MD system	[73]
48.4 g L^{-1} $MgCl_2$ & 49 g L^{-1} NaCl	Synthetic municipal wastewater	OMBR, CTA-ES (HTI,USA)	• Membrane fouling was not severe as shown by the decline in J_w with both NaCl and $MgCl_2$ • Accumulation of salts was observed with the use of both solutions	[24]
0.5 M NaCl & 0.35 M $MgCl_2$	Synthetic municipal wastewater	AnOMBR-MF system/ TFC (HTI,USA)	• Zero rejection of NH_4 –N using NaCl and 57.5–87.5% using $MgCl_2$ • $MgCl_2$ caused severe membrane inorganic fouling	[56]
1, 1.5 & 2 M $MgCl_2$	Municipal wastewater	Direct FO, TFC (Aquaporin A/S, Denmark)	• Both the increase in the $MgCl_2$ concentration and increase in the cross-flow rate contributed to the higher J_w, but COD concentration remained stable	[16]

Table 7.
Magnesium chloride as DS in FO treating municipal wastewater.

of the DS in the protocol. A bench-scale FO unit was used, while CTA membranes (Hydration Technology Innovations, HTI, USA) were applied to all experiments; the main purpose of this study was to evaluate the applicability of ionic organic solutions to OMBR systems.

Magnesium acetate ($C_4H_6MgO_4$) and sodium propionate ($C_3H_5NaO_2$) recorded the best performance as DS in terms of J_s, potential recovery, and biodegradability. Siddique et al. [76] showed similar results with the application of synthetic wastewater, highlighting $C_4H_6MgO_4$ as suitable DS for OMBR applications, while sodium acetate ($C_2H_3NaO_2$) led to the development of dense membrane biofilm. Despite the many benefits of ionic organic solutions, it should be noted that their potential application is limited, as the re-concentration cost is high compared to inorganic solutions.

A recent study aimed to integrate all the parameters of **Table 4** with the compatibility of FO as a pre-treatment step preceding the anaerobic process [83].

Draw	Feed	Configuration/ Membrane Type	Findings	Ref.
0.6 M $C_2H_3NaO_2$, 0.3 M EDTA-2Na & 0.5 M NaCl	Synthetic municipal wastewater	OMBR-RO/ CTA (HTI,USA)	• Minimized membrane fouling, lower J_s for the tested solutions over NaCl, but also lower J_w and reduced salt accumulation	[74]
0.1, 0.3, 0.5, 0.7 & 1 M EDTA-2Na	Activated sludge	FO-NF/ CTA membranes (HTI, USA)	• 0.7 M EDTA-2Na was the preferred concentration	[75]
			• At pH equal to 8, the FO membranes achieved the best performance	[75]
			• Reduced J_s compared to inorganic salts	
0.25 M $CaCl_2$, 0.25 M $MgCl_2$, 0.25 M $C_4H_6MgO_4$ & 0.25 M $C_2H_3NaO_2$	Synthetic municipal wastewater	OMBR-MD/CTA membrane (HTI, USA)	• $C_4H_6MgO_4$ optimized sludge flocculation, $C_2H_3NaO_2$ and $MgCl_2$ achieve steady flows in repeated tests	[76]
			• Inhibition of biological activity due to Cl^- presence	
			• $C_2H_3NaO_2$ generates significant fouling	
NaCl, $MgSO_4$, $C_2H_3NaO_2$, $C_4H_6MgO_4$, $CHNaO_2$, EDTA-2Na, $C_6H_{12}O_6$, $C_2H_5NO_2$, $C_3H_8O_3$, CH_4N_2O	Municipal wastewater	Direct FO/ CTA membrane (HTI)	• Organic solutions, such as $C_2H_3NaO_2$, are appropriate for this configuration. Reverse leakage of NaCl does not interrupt the anaerobic process	[13]
$CHKO_2$, $K_4P_2O_7$, $(C_3H_3NaO_2)_n$, $C_{2n}H_{4n} + 2O_{n+1}$, $MgSO_4$ & NaCl	Synthetic secondary effluent	Post FO/ spiral wound (SW) CTA and TFC & flat sheet TFC and CTA	• $MgSO_4$, $(C_3H_3NaO_2)_n$ and $K_4P_2O_7$ identified as suitable solutions in terms of cost-effectiveness, toxicity, recovery and pH range	[77]
$(C_3H_3NaO_2)_n$, $MgSO_4$ & $MgCl_2$	MBR permeate	FO-NF/ TFC (Porifera, CA, USA)	• $(C_3H_3NaO_2)_n$ was unsuitable for irrigation, $MgSO_4$ caused prolonged membrane fouling, while $MgCl_2$ had the best performance	[78]
Commercial fertiliser diamond blue	Raw wastewater, MBR supernatant & permeate	Direct FO/ TFC membrane (Toray Industry Inc.)	• Liquid fertilizer has a good performance comparable to the application with common inorganic salts for green wall irrigation	[79]
KNO_3, KH_2PO_4 &KNO_3 (fertilizers)	Synthetic secondary effluent	FDFO/ CTA membrane (HTI, USA)	• Occurrence of severe biofouling using KNO_3 as DS compared to KCl and KH_2PO_4; membrane flux decline by 63%, 45% and 30%, respectively	[80]

Draw	Feed	Configuration/ Membrane Type	Findings	Ref.
NH_4NO_3, $(NH_4)_2SO_4$, NH_4Cl, $Ca(NO_3)_2$, KCl, $NH_4H_2PO_4$, $(NH_4)_2HPO_4$, KNO_3 & KH_2PO_4	Synthetic municipal wastewater	FDFO/ TFC membrane, (Toray Industry Inc.)	• KCl and NH_4Cl showed the highest water recovery and MAP, KH_2PO_4 and SOA showed the lowest J_s • High dilution is required about 100 tunes	[55]
$(NH_4)_2SO_4$, KH_2PO_4 & $(NH_4)H_2PO_4$	Synthetic municipal wastewater	FO-MBR (Direct FO)/CTA-ES membrane (HTI)	• MAP had the best performance with the lowest J_s • SOA altered the sludge characteristics	[51]
Commercial liquid fertilizer	Raw wastewater	Direct FO, CTA membrane (HTI, USA)	• Effective application of liquid fertilizer as DS for green wall irrigation • Reverse nutrient leakage worsened as the temperature rose	[81]
KH_2PO_4, KCl & KNO_3	Synthetic municipal wastewater	FDFO-AnMBR	• KH_2PO_4 < KCl < KNO_3 in terms of reverse leakages; alterations observed in anaerobic biomass, especially using KNO_3	[82]

Table 8.
Fertilizers, organic, inorganic, and ionic organic compounds as DS in FO treating municipal wastewater.

Among the 5 different zwitterions solutions tested, glycine ($C_2H_5NO_2$), L-proline ($C_5H_9NO_2$), and glycine betaine ($C_5H_{11}NO_2$) exhibited comparable J_w to NaCl (4.3–4.9 L m^{-2} h^{-1}), with lower J_s. From a physicochemical perspective, the process efficiency depends significantly on the pH value, affecting both the charge and the molecular size. Despite the rapid biodegradation (Adenosine triphosphate (ATP) levels range from 7 to 14 μg L^{-1} after degradation tests) of all zwitterions compounds, the replacement cost, which is 3–4 times more than the cost of commercially available solutions, is a potential barrier to their implementation in municipal wastewater streams. It is worth noting that the above experiments were performed with deionized water as feed, which favors the overall performance over the application of a more complex ionic matrix, such as sewage [83].

Commercial fertilizers are another largely inorganic solution medium that has been tested in various effluents resulting from a WWTP, such as typical secondary and MBR permeate and raw municipal wastewater. As illustrated in **Table 8**. Li et al. [82] compared the effect of 3 different commercial fertilizers on the downstream anaerobic process when applied as draw agents directly in raw wastewater. The following order of compatibility with the anaerobic treatment revealed Potassium Nitrate (KNO_3) > Potassium Chloride (KCl) > Potassium dihydrogen Phosphate (KH_2PO_4), with their reverse solute flux showing a similar sequence when the concentration of all DS was equal to 1 M. Water flux can be dramatically reduced by applying KNO_3 as DS, as extensive biofouling has been observed, while increasing nitrate concentrations can inhibit the subsequent anaerobic process, rendering them as unsuitable [80]. The implementation of different fertilizers in a hybrid FO-RO system to concentrate MBR permeate proved that the amplification of enhanced NaCl with Diammonium Phosphate (DAP) ($(NH_4)_2HPO_4$) can reduce reverse solute leakage by 35%, achieving NH_4-N rejection rates more than 95% at different flow rates (1.2 and 2 L m^{-2} h^{-1}) [66]. In addition, a long-term study examining the pilot application of a hybrid FO-NF system that treated MBR permeate found that Sodium Polyacrylate ($(C_3H_3NaO_2)_n$) was inappropriate for irrigation practices. On the contrary, the combination of $MgCl_2$ with NF membranes significantly improved the process efficiency and operating costs, as the application of chemical cleaning was not required. However, a notably high loss of the osmotic agent was observed [78]. A particularly interesting investigation was carried out by Adnan et al. [51] in which the possibility of applying 9 different fertilizers to the direct FO for the wastewater valorization and its parallel application in agricultural practices was examined. Water recovery was high by applying KCl (J_w = 21.1 L m^{-2} h^{-1}; J_s = 11.2 g m^{-2} h^{-1}; Osmotic Pressure (OP) = 44.6 bar) and Ammonium Chloride (NH_4Cl) (J_w = 21.1 L m^{-2} h^{-1}; J_s = 7.5 g m^{-2} h^{-1}; OP = 43.5 bar), while other fertilizers recorded particularly low reverse flux, such as Ammonium Sulfate (SOA) ($(NH_4)_2SO_4$) (J_w = 15.5 L m^{-2} h^{-1}; J_s = 1.7 g m^{-2} h^{-1}; OP = 46.7 bar), KH_2PO_4 (J_w = 13.2 L m^{-2} h^{-1}; J_s = 2.3 g m^{-2} h^{-1}; OP = 36.5 bar), and $NH_4H_2PO_4$ (Monoammonium Phosphate, MAP) (J_w = 13.8 L m^{-2} h^{-1}; J_s = 1 g m^{-2} h^{-1}; OP = 44.4 bar). However, this process becomes inapplicable, as a large amount of water is required to dilute the concentrated fertilizer (at least 1/100), to reach the irrigation limits [51].

The analysis of the existing literature makes it clear that the FO process is still under investigation and the determination of the ideal DS plays a vital role in upgrading the process of this technology. Despite the properties of the DS, the selection of the suitable configuration, the techno-economic factors, and the re-condensation method should be combined during the selection process; the optimization of the FO membrane's properties is a major challenge that can solve many issues. The development and fabrication of higher rejection membranes can be the answer to the implementation of both monovalent and divalent ions, which have been widely used as DS and their performance is already known to the research community.

4. Integration of FO followed by anaerobic treatment in a WWTP

4.1 COD valorization in municipal WWTPs

For more than a century, the CAS process has been applied as the main urban wastewater treatment system worldwide, making a significant contribution to environmental protection and public health. However, the low energy efficiency of the CAS process ranks WWTPs among the largest energy consumers in a country; on an annual basis, in developed counties, about 1–3% of electricity consumption is spent on their operation [84]. In addition, WWTPs are characterized by a high energy and carbon footprint, as during biological processes, large amounts of greenhouse gases are produced, mainly carbon dioxide generated due to the oxidation of organic matter and indirectly by electricity consumption [85]. Therefore, about 0.3–0.5 kWh m^{-3} of energy is required for sewage treatment by applying the CAS process, while the contained chemical energy and nutrients are not utilized [86].

According to Wan et al. [87] the traditional CAS process needs an average of 0.45 kWh to treat one m^3 of sewage, which equals to 1620 kJ m^{-3}. Assuming a concentration of 600 mg L^{-1} COD, energy consumption becomes 2.7 kJ g^{-1} COD.

As shown in **Figure 2(A)**, the energy recovery in convectional CAS systems occurs through the anaerobic digestion of the primary and secondary sludge, which corresponds to 32–39% of the organic material in COD terms. The latter percentage is equal to 2.9–3.5 kJ g^{-1} COD, since 1 g of methane-COD is equal to 13.9 kJ (65% methane percent in produced biogas). Considering that only 35% of the produced methane can be utilized for the production of electricity [86], about 1–1.2 kJ g^{-1} COD can be recovered from municipal wastewater, by applying anaerobic digestion to the sludge treatment line. Comparing the aforementioned energy requirement, 2.7 kJ g^{-1} COD, it is estimated that about 40% of it can be recovered using anaerobic digestion (1–1.2 kJ g^{-1} COD). The anaerobic digestion process also generates approximately 50–55% heat, part of which is used to heat the digesters. The excess heat can only be valorized locally [88].

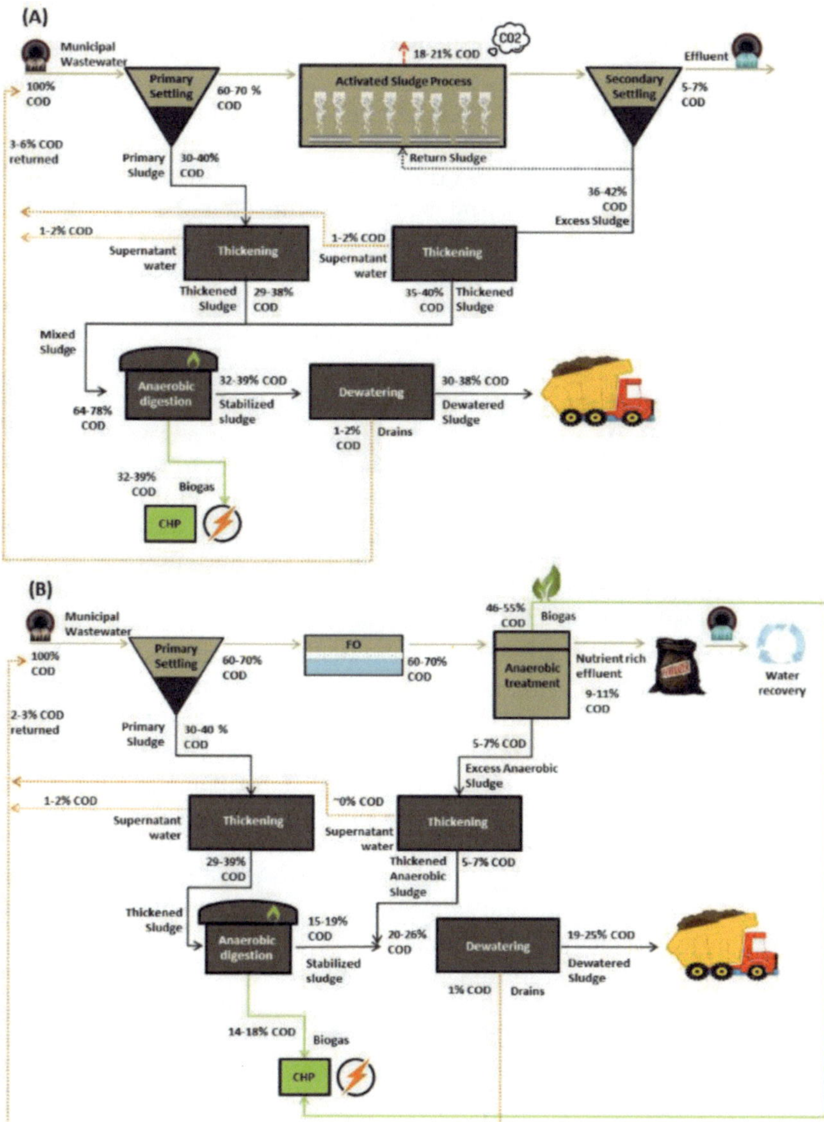

Figure 2.
(A) COD mass flow in a convectional WWTP, (B) COD mass flow, when FO followed by anaerobic treatment is integrated into a WWTP.

Obviously, COD capture, and subsequently valorization of the chemical energy contained in municipal wastewater can lead WWTPs to sustainable development, transforming WWTPs from energy consumers to producers, while significantly reducing the environmental footprint and operating costs.

The integration of FO in municipal wastewater treatment and the benefits of its application have been investigated in various studies [17, 49]. This chapter presents the combination of FO and anaerobic treatment in a typical WWTP for the utilization of the chemical energy, which is inherently present in sewage. As shown in **Figure 2(B)**, by placing the FO in the main treatment line of a WWTP and taking into account the efficiency of a typical anaerobic system, such as An-MBR, which is equal to 80% in ambient conditions [89], 46–55% of COD is converted to biogas (65% of the aforementioned percent corresponds to methane). Following the same procedure as before, the energy recovery in the main treatment line through the implementation of anaerobic treatment is between 4.2–5 kJ g^{-1} COD. Another 1.3–1.6 kJ g^{-1} COD of energy is recovered from the anaerobic digestion of the sewage sludge (13.9 kJ g^{-1} methane-COD). Since only 35% of the produced methane can be converted into electricity [86], the power production from the wastewater treatment line ranges between 1.3–1.7 kJ g^{-1} COD, while from the sludge treatment line it is equal to 0.4–0.6 kJ g^{-1} COD. On aggregate, 1.9–2.3 kJ g^{-1} COD of electricity can be utilized from this innovative treatment scheme, which can counterbalance 80% of the existing energy consumption of a typical municipal WWTP. The treated effluent of the anaerobic system is rich in nutrients, which can be valorized by applying recovery technologies for the production of slow-release fertilizers, while the reclaimed water content can also be reused.

4.2 Salinity, the greatest impact of FO on anaerobic treatment

Despite the benefits of the wastewater management system presented in the above section, there are two factors that can be particularly limiting to the subsequent operation of the anaerobic process. The solute flux that characterized the FO system results in the accumulation of salts in the feed stream, potentially resulting in partial or complete inhibition of the downstream anaerobic and aerobic biological treatment processes [14, 17, 32]. Salinity has been identified as an inhibitory agent of the anaerobic process, as the increased osmotic pressure across the cell membrane can cause plasmolysis, leading to cell death and total inhibition of the anaerobic process. More specifically, Lefebvre et al. [90] stressed that the activity of methanogenic bacteria is inhibited at concentrations of NaCl equal to 5 g L^{-1}, while acidogenic microorganisms are affected at much higher concentrations, i.e. 20 g L^{-1}. Ansari et al. [91] studied the effects of NaCl on anaerobic treatment of concentrated wastewater effluents in batch mode experiments and observed that by increasing water recovery rates of FO (from 50 to 90%), the anaerobic process achieves higher methane production (approximately 5 times higher), while the presence of salinity has a negligible negative effect.

Based on the existing literature, the limiting parameter of salinity has been investigated and observed only in aerobic/anaerobic systems, where the FO unit is plugged into MBR systems for a relatively short time, while in pre-concentration systems few studies have examined the effect of salinity on the downstream anaerobic process and suggest mitigation measures. Chen et al. [37] and Wang et al. [39] did not observe significant effects of salinity on anaerobic reactors by recording an average methane yield of 0.2 and 0.3 L CH_4 g^{-1} COD, respectively, in studies that cannot be characterized as long-term. As mentioned above, the application of minimization strategies such as the corresponding regulation of the hydraulic residence time (HRT) seems to regulate the salinity conditions to which the biomass is exposed. Accordingly, the addition of MF membranes is a particular interesting approach for the minimization of salinity and the parallel application of nutrient recovery methods. Another interesting perspective is the acclimatization of the anaerobic biomass to high salinity conditions. This mitigation technique is not recent as the presence of specific microorganisms, such as halotolerant bacteria has shown particularly high efficiency in the anaerobic treatment of saline industrial wastewater [92]. In a recent study, where no acclimatized biomass was used, Gao et al. [93] separately investigated the effect of high salinity and ammonia nitrogen concentration and the combination of the two inhibitors in the anaerobic treatment of pre-concentrated municipal wastewater. The results showed that the presence of NH_4-N and NaCl concentrations separately, up to 200 mg L^{-1} and between 5 and 8 g L^{-1}, respectively, did not significantly affect the activity of anaerobic microorganisms. The combination of the two parameters in non-acclimatized and acclimatized biomass showed that the latter had significantly better performance and can respond without the risk of inhibition. Further research into anaerobic biomass acclimatization should be conducted in the future, as higher condensation rates could be applied from the upstream FO unit.

All the acquired knowledge of the above studies would be particularly interesting to be used in the long-term investigation of a FO system combined with a downstream anaerobic process, in which all the limiting parameters and the proposed mitigation measures can be examined in-depth, for the rational assessment of its performance.

5. Conclusion

There is no doubt that FO is a promising technology that has been investigated for a range of applications at various stages of a municipal WWTP. Among them, its combination with the anaerobic process has significant advantages, as much of the chemical energy inherently contained in sewage can be recovered as biogas, while resource recovery technologies can be applied downstream, utilizing the nutrient-rich effluent. However, the transition of the FO from laboratory scale to full-scale applications requires further research to address important issues, such as the salinity accumulation in the downstream technologies and the reduced rejection of NH_4-N by existing FO membranes. The application of NaCl indicates a possible suitability for the concentration of municipal wastewater. The background knowledge available on the basic criteria of FO has to be utilized for the development of membranes with higher selectivity. Future investigations should carry out extensive long-term monitoring and targeted combination/interaction of different parameters for the concentration of real wastewater, to assess from a technical, environmental and economic perspective the feasibility of applying FO technology to municipal wastewater management.

Acknowledgements

This research work was supported by the Hellenic Foundation for Research and Innovation (H.F.R.I.) under the “First Call for H.F.R.I. Research Projects to support Faculty members and Researchers and the procurement of high-cost research equipment grant” (Project Number: 4008).

Nomenclature

ATP	adenosine triphosphate
NH_4-N	ammonium nitrogen
SOA	ammonium sulfate
An-OMBR	anaerobic osmotic membrane bioreactor
CTA	cellulose triacetate
COD	chemical oxygen demand
CAS	conventional activated sludge
DAP	diammonium phosphate
DS	draw solution
EDTA 2Na	ethylenediamine tetraacetic acid disodium salt
EU	European Union
FO	forward osmosis
FO-NF	forward osmosis-nanofiltration
$C_2H_5NO_2$	glycine
$C_5H_{11}NO_2$	glycine betaine
HTI	hydration technology innovations
HRT	hydraulic residence time
$C_5H_9NO_2$	L-proline
$C_4H_6MgO_4$	magnesium acetate
$MgCl_2$	magnesium chloride
$MgSO_4$	magnesium sulfate
MBR	membrane bioreactor
MF	microfiltration
MF-OMBR	microfiltration-osmotic membrane bioreactor
MAP	monoammonium phosphate
NF	nanofiltration
PO_4-P	orthophosphate as phosphorus
OMBR	osmotic membrane bioreactor
KCl	potassium chloride
KH_2PO_4	potassium dihydrogen phosphate
KNO_3	potassium nitrate
RO	reverse osmosis
J_s	reverse salt flux
SRT	sludge retention time
$C_2H_3NaO_2$	sodium acetate
NaCl	sodium chloride
$(C_3H_3NaO_2)_n$	sodium polyacrylate
$C_3H_5NaO_2$	sodium propionate
TFC	thin film composite
TOC	total organic cabon
UF	ultrafiltration
WWTP(s)	wastewater treatment plant(s)
J_w	water flux

Author details

Stavroula Kappa* and Simos Malamis
Department of Water Resources and Environmental Engineering, School of Civil Engineering, National Technical University of Athens, Athens, Greece

*Address all correspondence to: stavroula_kappa@windowslive.com

IntechOpen

© 2021 The Author(s). Licensee IntechOpen. This chapter is distributed under the terms of the Creative Commons Attribution License (http://creativecommons.org/licenses/by/3.0), which permits unrestricted use, distribution, and reproduction in any medium, provided the original work is properly cited.

References

[1] Bond NR, Burrows RM, Kennard MJ, Bunn SE. Water scarcity as a driver of multiple stressor effects. Elsevier Inc.; 2018. DOI:10.1016/B978-0-12-811713-2.00006-6.

[2] Verner D. Climate Change Contributes to Water Scarcity. In: Verner D, editor. Adaptation to a Changing Climate in the Arab Countries, 2012, p. 110–51. DOI: 10.1596/978-0-8213-9459-5.

[3] European Commission (December 2nd 2015). Closing the loop—An EU Action Plan for the Circular Economy. Communication from the Commission to the European Parliament, the Council, the European Economic and Social Committee and the Committee of the Regions COM (2015), 614, Brussels. n.d.

[4] Escobar IC. Chapter 14 Conclusion: A Summary of Challenges still Facing Desalination and Water Reuse. vol. 2. Elsevier; 2010. DOI:10.1016/S1871-2711(09)00214-1.

[5] Ge Q, Ling M, Chung TS. Draw solutions for forward osmosis processes: Developments, challenges, and prospects for the future. Journal of Membrane Science. 2013;442:225–37. DOI:10.1016/j.memsci.2013.03.046.

[6] Lee KP, Arnot TC, Mattia D. A review of reverse osmosis membrane materials for desalination-Development to date and future potential. Journal of Membrane Science. 2011;370:1–22. DOI: 10.1016/j.memsci.2010.12.036.

[7] Sun Y, Tian J, Zhao Z, Shi W, Liu D, Cui F. Membrane fouling of forward osmosis (FO) membrane for municipal wastewater treatment: A comparison between direct FO and OMBR. Water Research. 2016;104:330–9. DOI:10.1016/j.watres.2016.08.039.

[8] Zhao S, Zou L, Tang CY, Mulcahy D. Recent developments in forward osmosis: Opportunities and challenges. Journal of Membrane Science. 2012; 396:1–21. DOI:10.1016/j.memsci.2011.12.023.

[9] Chung TS, Li X, Ong RC, Ge Q, Wang H, Han G. Emerging forward osmosis (FO) technologies and challenges ahead for clean water and clean energy applications. Current Opinion in Chemical Engineering. 2012; 1:246–57. DOI:10.1016/j.coche.2012.07.004.

[10] Ang WL, Wahab Mohammad A, Johnson D, Hilal N. Forward osmosis research trends in desalination and wastewater treatment: A review of research trends over the past decade. Journal of Water Process Engineering. 2019;31:100886. DOI:10.1016/j.jwpe.2019.100886.

[11] Ansari AJ, Hai FI, Guo W, Ngo HH, Price WE, Nghiem LD. Factors governing the pre-concentration of wastewater using forward osmosis for subsequent resource recovery. Science of the Total Environment. 2016;566–567:559–66. DOI:10.1016/j.scitotenv.2016.05.139.

[12] Qiu G, Law YM, Das S, Ting YP. Direct and complete phosphorus recovery from municipal wastewater using a hybrid microfiltration-forward osmosis membrane bioreactor process with seawater brine as draw solution. Environmental Science and Technology. 2015;49:6156–63. DOI:10.1021/es504554f.

[13] Ansari AJ, Hai FI, Guo W, Ngo HH, Price WE, Nghiem LD. Selection of forward osmosis draw solutes for subsequent integration with anaerobic treatment to facilitate resource recovery from wastewater. Bioresource Technology. 2015;191:30–6. DOI: 10.1016/j.biortech.2015.04.119.

[14] Bowden KS, Achilli A, Childress AE. Organic ionic salt draw solutions for osmotic membrane bioreactors. Bioresource Technology. 2012;122:207–16. DOI:10.1016/j.biortech.2012.06.026.

[15] Achilli A, Cath TY, Childress AE. Selection of inorganic-based draw solutions for forward osmosis applications. Journal of Membrane Science. 2010;364:233–41. DOI:10.1016/j.memsci.2010.08.010.

[16] Singh N, Dhiman S, Basu S, Balakrishnan M, Petrinic I, Helix-Nielsen C. Dewatering of sewage for nutrients and water recovery by Forward Osmosis (FO)using divalent draw solution. Journal of Water Process Engineering. 2019;31:100853. DOI: 10.1016/j.jwpe.2019.100853.

[17] Wang Z, Zheng J, Tang J, Wang X, Wu Z. A pilot-scale forward osmosis membrane system for concentrating low-strength municipal wastewater: Performance and implications. Scientific Reports. 2016;6:1–11. DOI: 10.1038/srep21653.

[18] Korenak J, Basu S, Balakrishnan M, Hélix-Nielsen C, Petrinic I. Forward osmosis in wastewater treatment processes. Acta Chimica Slovenica. 2017;64:83–94. DOI:10.17344/acsi.2016.2852.

[19] Cornelissen ER, Harmsen D, de Korte KF, Ruiken CJ, Qin JJ, Oo H, Wessels LP. Membrane fouling and process performance of forward osmosis membranes on activated sludge. Journal of Membrane Science. 2008;319:158–68. DOI:10.1016/j.memsci.2008.03.048.

[20] Wang X, Zhao Y, Yuan B, Wang Z, Li X, Ren Y. Comparison of biofouling mechanisms between cellulose triacetate (CTA) and thin-film composite (TFC) polyamide forward osmosis membranes in osmotic membrane bioreactors. Bioresource Technology. 2016;202:50–8. DOI:10.1016/j.biortech.2015.11.087.

[21] Holloway RW, Regnery J, Nghiem LD, Cath TY. Removal of trace organic chemicals and performance of a novel hybrid ultrafiltration-osmotic membrane bioreactor. Environmental Science and Technology. 2014;48: 10859–68. DOI:10.1021/es501051b.

[22] Qiu G, Ting YP. Osmotic membrane bioreactor for wastewater treatment and the effect of salt accumulation on system performance and microbial community dynamics. Bioresource Technology. 2013;150:287–97. DOI: 10.1016/j.biortech.2013.09.090.

[23] Sun Y, Tian J, Song L, Gao S, Shi W, Cui F. Dynamic changes of the fouling layer in forward osmosis based membrane processes for municipal wastewater treatment. Journal of Membrane Science. 2018;549:523–32. DOI:10.1016/j.memsci.2017.12.055.

[24] Qiu G, Ting YP. Direct phosphorus recovery from municipal wastewater via osmotic membrane bioreactor (OMBR) for wastewater treatment. Bioresource Technology. 2014;170: 221–9. DOI:10.1016/j.biortech.2014.07.103.

[25] Huang L ying, Lee DJ, Lai JY. Forward osmosis membrane bioreactor for wastewater treatment with phosphorus recovery. Bioresource Technology. 2015;198:418–23. DOI: 10.1016/j.biortech.2015.09.045.

[26] Luo W, Hai FI, Price WE, Guo W, Ngo HH, Yamamoto K, Nghiem LD. Phosphorus and water recovery by a novel osmotic membrane bioreactor-reverse osmosis system. Bioresource Technology. 2016;200:297–304. DOI: 10.1016/j.biortech.2015.10.029.

[27] Luo W, Xie M, Song X, Guo W, Ngo HH, Zhou JL, Nghiem LD. Biomimetic aquaporin membranes for osmotic membrane bioreactors: Membrane performance and contaminant removal. Bioresource

Technology. 2018;249:62–8. DOI: 10.1016/j.biortech.2017.09.170.

[28] Chang HM, Chen SS, Nguyen NC, Chang WS, Sinha Ray S. Osmosis membrane bioreactor–microfiltration with magnesium-based draw solute for salinity reduction and phosphorus recovery. International Biodeterioration and Biodegradation. 2017;124:169–75. DOI:10.1016/j.ibiod.2017.05.022.

[29] Xu Z, Song X, Xie M, Wang Y, Huda N, Li G, Luo W. Effects of surfactant addition to draw solution on the performance of osmotic membrane bioreactor. Journal of Membrane Science. 2021;618:118634. DOI:10.1016/j.memsci.2020.118634.

[30] Ab Hamid NH, Wang DK, Smart S, Ye L. Achieving stable operation and shortcut nitrogen removal in a long-term operated aerobic forward osmosis membrane bioreactor (FOMBR) for treating municipal wastewater. Chemosphere. 2020;260. DOI:10.1016/j.chemosphere.2020.127581.

[31] Yap WJ, Zhang J, Lay WCL, Cao B, Fane AG, Liu Y. State of the art of osmotic membrane bioreactors for water reclamation. Bioresource Technology. 2012;122:217–22. DOI: 10.1016/j.biortech.2012.03.060.

[32] Wang X, Chang VWC, Tang CY. Osmotic membrane bioreactor (OMBR) technology for wastewater treatment and reclamation: Advances, challenges, and prospects for the future. Journal of Membrane Science. 2016;504:113–32. DOI:10.1016/j.memsci.2016.01.010.

[33] Christian S, Grant S, McCarthy P, Wilson D, Mills D. The first two years of full-scale anaerobic membrane bioreactor (AnMBR) operation treating high-strength industrial wastewater. Water Practice and Technology. 2011;6. DOI:10.2166/wpt.2011.032.

[34] Ansari AJ, Hai FI, Price WE, Drewes JE, Nghiem LD. Forward osmosis as a platform for resource recovery from municipal wastewater - A critical assessment of the literature. Journal of Membrane Science. 2017;529: 195–206. DOI:10.1016/j.memsci.2017.01.054.

[35] Zhang H, Jiang W, Cui H. Performance of anaerobic forward osmosis membrane bioreactor coupled with microbial electrolysis cell (AnOMEBR) for energy recovery and membrane fouling alleviation. Chemical Engineering Journal. 2017;321:375–83. DOI:10.1016/j.cej.2017.03.134.

[36] Gu Y, Chen L, Ng JW, Lee C, Chang VWC, Tang CY. Development of anaerobic osmotic membrane bioreactor for low-strength wastewater treatment at mesophilic condition. Journal of Membrane Science. 2015;490:197–208. DOI:10.1016/j.memsci.2015.04.032.

[37] Chen L, Gu Y, Cao C, Zhang J, Ng JW, Tang C. Performance of a submerged anaerobic membrane bioreactor with forward osmosis membrane for low-strength wastewater treatment. Water Research. 2014;50:114–23. DOI:10.1016/j.watres.2013.12.009.

[38] Chang HM, Sun YC, Chien IC, Chang WS, Ray SS, Cao DTN, Cong Duong C, Chen SS. Innovative upflow anaerobic sludge osmotic membrane bioreactor for wastewater treatment. Bioresource Technology. 2019;287: 121466. DOI:10.1016/j.biortech.2019.121466.

[39] Wang X, Wang C, Tang CY, Hu T, Li X, Ren Y. Development of a novel anaerobic membrane bioreactor simultaneously integrating microfiltration and forward osmosis membranes for low-strength wastewater treatment. Journal of Membrane Science. 2017;527:1–7. DOI: 10.1016/j.memsci.2016.12.062.

[40] Gao Y, Fang Z, Chen C, Zhu X, Liang P, Qiu Y, Zhang X, Huang X.

Evaluating the performance of inorganic draw solution concentrations in an anaerobic forward osmosis membrane bioreactor for real municipal sewage treatment. Bioresource Technology. 2020;307:123254. DOI:10.1016/j.biortech.2020.123254.

[41] Hou D, Lu L, Ren ZJ. Microbial fuel cells and osmotic membrane bioreactors have mutual benefits for wastewater treatment and energy production. Water Research. 2016;98:183–9. DOI: 10.1016/j.watres.2016.04.017.

[42] Hou D, Lu L, Sun D, Ge Z, Huang X, Cath TY, Ren ZJ. Microbial electrochemical nutrient recovery in anaerobic osmotic membrane bioreactors. Water Research. 2017;114: 181–8. DOI:10.1016/j.watres.2017.02.034.

[43] Holloway RW, Wait AS, Fernandes da Silva A, Herron J, Schutter MD, Lampi K, Cath TY. Long-term pilot scale investigation of novel hybrid ultrafiltration-osmotic membrane bioreactors. Desalination. 2015;363:64–74. DOI:10.1016/j.desal.2014.05.040.

[44] Luo W, Xie M, Hai FI, Price WE, Nghiem LD. Biodegradation of cellulose triacetate and polyamide forward osmosis membranes in an activated sludge bioreactor: Observations and implications. Journal of Membrane Science. 2016;510:284–92. DOI:10.1016/j.memsci.2016.02.066.

[45] Zhang X, Ning Z, Wang DK, Diniz da Costa JC. Processing municipal wastewaters by forward osmosis using CTA membrane. Journal of Membrane Science. 2014;468:269–75. DOI:10.1016/j.memsci.2014.06.016.

[46] Gao Y, Fang Z, Liang P, Huang X. Direct concentration of municipal sewage by forward osmosis and membrane fouling behavior. Bioresource Technology. 2018;247:730–5. DOI:10.1016/j.biortech.2017.09.145.

[47] Ferrari F, Pijuan M, Rodriguez-Roda I, Blandin G. Exploring submerged forward osmosis for water recovery and pre-concentration of wastewater before anaerobic digestion: A pilot scale study. Membranes. 2019;9. DOI:10.3390/membranes9080097.

[48] Rood B, Zhang C, Inniss E, Hu Z. Forward osmosis with an algal draw solution to concentrate municipal wastewater and recover resources. Water Environment Research. 2020;92: 689–97. DOI:10.1002/wer.1262.

[49] Yang S, Gao B, Jang A, Shon H kyong, Yue Q. Municipal wastewater treatment by forward osmosis using seawater concentrate as draw solution. Chemosphere. 2019;237:124485. DOI: 10.1016/j.chemosphere.2019.124485.

[50] Valladares Linares R, Li Z, Abu-Ghdaib M, Wei CH, Amy G, Vrouwenvelder JS. Water harvesting from municipal wastewater via osmotic gradient: An evaluation of process performance. Journal of Membrane Science. 2013;447:50–6. DOI:10.1016/j.memsci.2013.07.018.

[51] Adnan M, Khan SJ, Manzoor K, Hankins NP. Performance evaluation of fertilizer draw solutions for forward osmosis membrane bioreactor treating domestic wastewater. Process Safety and Environmental Protection. 2019; 127:133–40. DOI:10.1016/j.psep.2019.05.006.

[52] Blandin G, Ferrari F, Lesage G, Le-Clech P, Héran M, Martinez-Lladó X. Forward osmosis as concentration process: Review of opportunities and challenges. Membranes. 2020;10: 1–40. DOI:10.3390/membranes10100284.

[53] Chekli L, Phuntsho S, Shon HK, Vigneswaran S, Kandasamy J, Chanan A. A review of draw solutes in forward osmosis process and their use in modern applications. Desalination and

Water Treatment. 2012;43:167–84. DOI: 10.1080/19443994.2012.672168.

[54] Johnson DJ, Suwaileh WA, Mohammed AW, Hilal N. Osmotic ' s potential : An overview of draw solutes for forward osmosis. Desalination. 2018; 434:100–20.

[55] Chekli L, Kim Y, Phuntsho S, Li S, Ghaffour N, Leiknes TO, Shon HK. Evaluation of fertilizer-drawn forward osmosis for sustainable agriculture and water reuse in arid regions. Journal of Environmental Management. 2017;187: 137–45. DOI:10.1016/j.jenvman. 2016.11.021.

[56] Hu T, Wang X, Wang C, Li X, Ren Y. Impacts of inorganic draw solutes on the performance of thin-film composite forward osmosis membrane in a microfiltration assisted anaerobic osmotic membrane bioreactor. RSC Advances. 2017;7:16057–63. DOI: 10.1039/C7RA01524K.

[57] Long Q, Jia Y, Li J, Yang J, Liu F, Zheng J, Yu B. Recent advance on draw solutes development in Forward osmosis. Processes. 2018;6:7–11. DOI: 10.3390/pr6090165.

[58] Lutchmiah K, Verliefde ARD, Roest K, Rietveld LC, Cornelissen ER. Forward osmosis for application in wastewater treatment: A review. Water Research. 2014;58:179–97. DOI:10.1016/ j.watres.2014.03.045.

[59] Aftab B, Khan SJ, Maqbool T, Hankins NP. High strength domestic wastewater treatment with submerged forward osmosis membrane bioreactor. Water Science and Technology. 2015;72:141–8. DOI: 10.2166/wst.2015.195.

[60] Li J, Hou D, Li K, Zhang Y, Wang J, Zhang X. Domestic wastewater treatment by forward osmosis-membrane distillation (FO-MD) integrated system. Water Science and Technology. 2018;77:1514–23. DOI: 10.2166/wst.2018.031.

[61] Motsa MM, Mamba BB, Verliefde ARD. Forward osmosis membrane performance during simulated wastewater reclamation: Fouling mechanisms and fouling layer properties. Journal of Water Process Engineering. 2018;23:109–18. DOI: 10.1016/j.jwpe.2018.03.007.

[62] Jiang W, Lin L, Gedara SMH, Schaub TM, Jarvis JM, Wang X, Xu X, Nirmalakhandan N, Xu P. Potable-quality water recovery from primary effluent through a coupled algal-osmosis membrane system. Chemosphere. 2020; 240:124883. DOI:10.1016/j. chemosphere.2019.124883.

[63] Kedwell KC, Christensen ML, Quist-Jensen CA, Jørgensen MK. Effect of reverse sodium flux and pH on ammoniacal nitrogen transport through biomimetic membranes. Separation and Purification Technology. 2019;217:40–7. DOI:10.1016/j.seppur.2019.02.001.

[64] Yin Tang MK, Ng HY. Impacts of different draw solutions on a novel anaerobic forward osmosis membrane bioreactor (AnFOMBR). Water Science and Technology. 2014;69:2036–42. DOI: 10.2166/wst.2014.116.

[65] Luján-Facundo MJ, Soler-Cabezas JL, Mendoza-Roca JA, Vincent-Vela MC, Bes-Piá A, Doñate-Hernández S. A study of the osmotic membrane bioreactor process using a sodium chloride solution and an industrial effluent as draw solutions. Chemical Engineering Journal. 2017;322:603–10. DOI:10.1016/j.cej.2017.04.062.

[66] Hafiz MA, Hawari AH, Altaee A. A hybrid forward osmosis/reverse osmosis process for the supply of fertilizing solution from treated wastewater. Journal of Water Process Engineering. 2019;32. DOI:10.1016/j.jwpe.2019. 100975.

[67] Xue W, Tobino T, Nakajima F, Yamamoto K. Seawater-driven forward osmosis for enriching nitrogen and phosphorous in treated municipal wastewater: Effect of membrane properties and feed solution chemistry. Water Research. 2015;69:120–30. DOI: 10.1016/j.watres.2014.11.007.

[68] Soler-Cabezas JL, Mendoza-Roca JA, Vincent-Vela MC, Luján-Facundo MJ, Pastor-Alcañiz L. Simultaneous concentration of nutrients from anaerobically digested sludge centrate and pre-treatment of industrial effluents by forward osmosis. Separation and Purification Technology. 2018;193:289–96. DOI:10.1016/j.seppur.2017.10.058.

[69] Vu MT, Price WE, He T, Zhang X, Nghiem LD. Seawater-driven forward osmosis for pre-concentrating nutrients in digested sludge centrate. Journal of Environmental Management. 2019;247: 135–9. DOI:10.1016/j.jenvman.2019.06.082.

[70] Blandin G, Rosselló B, Monsalvo VM, Batlle-Vilanova P, Viñas JM, Rogalla F, Comas J. Volatile fatty acids concentration in real wastewater by forward osmosis. Journal of Membrane Science. 2019;575: 60–70. DOI:10.1016/j.memsci.2019.01.006.

[71] Devia YP, Imai T, Higuchi T, Kanno A, Yamamoto K, Sekine M, Le T Van. Potential of Magnesium Chloride for Nutrient Rejection in Forward Osmosis. Journal of Water Resource and Protection. 2015;07:730–40. DOI: 10.4236/jwarp.2015.79060.

[72] Nguyen NC, Chen SS, Nguyen HT, Ray SS, Ngo HH, Guo W, Lin PH. Innovative sponge-based moving bed-osmotic membrane bioreactor hybrid system using a new class of draw solution for municipal wastewater treatment. Water Research. 2016;91: 305–13. DOI:10.1016/j.watres.2016.01.024.

[73] Xie M, Nghiem LD, Price WE, Elimelech M. Toward Resource Recovery from Wastewater: Extraction of Phosphorus from Digested Sludge Using a Hybrid Forward Osmosis-Membrane Distillation Process. Environmental Science and Technology Letters. 2014;1:191–5. DOI:10.1021/ez400189z.

[74] Luo W, Hai FI, Price WE, Elimelech M, Nghiem LD. Evaluating ionic organic draw solutes in osmotic membrane bioreactors for water reuse. Journal of Membrane Science. 2016;514: 636–45. DOI:10.1016/j.memsci.2016.05.023.

[75] Hau NT, Chen SS, Nguyen NC, Huang KZ, Ngo HH, Guo W. Exploration of EDTA sodium salt as novel draw solution in forward osmosis process for dewatering of high nutrient sludge. Journal of Membrane Science. 2014;455:305–11. DOI:10.1016/j.memsci.2013.12.068.

[76] Siddique MS, Khan SJ, Shahzad MA, Nawaz MS, Hankins NP. Insight into the effect of organic and inorganic draw solutes on the flux stability and sludge characteristics in the osmotic membrane bioreactor. Bioresource Technology. 2018;249:758–66. DOI:10.1016/j.biortech.2017.10.064.

[77] Corzo B, de la Torre T, Sans C, Ferrero E, Malfeito JJ. Evaluation of draw solutions and commercially available forward osmosis membrane modules for wastewater reclamation at pilot scale. Chemical Engineering Journal. 2017;326:1–8. DOI:10.1016/j.cej.2017.05.108.

[78] Corzo B, de la Torre T, Sans C, Escorihuela R, Navea S, Malfeito JJ. Long-term evaluation of a forward osmosis-nanofiltration demonstration plant for wastewater reuse in agriculture. Chemical Engineering Journal. 2018;338:383–91. DOI:10.1016/j.cej.2018.01.042.

[79] Kim JE, Kuntz J, Jang A, Kim IS, Choi JY, Phuntsho S, Shon HK. Techno-economic assessment of fertiliser drawn forward osmosis process for greenwall plants from urban wastewater. Process Safety and Environmental Protection. 2019;127:180–8. DOI:10.1016/j.psep.2019.05.014.

[80] Li S, Kim Y, Chekli L, Phuntsho S, Shon HK, Leiknes TO, Ghaffour N. Impact of reverse nutrient diffusion on membrane biofouling in fertilizer-drawn forward osmosis. Journal of Membrane Science. 2017;539:108–15. DOI:10.1016/j.memsci.2017.05.074.

[81] Xie M, Zheng M, Cooper P, Price WE, Nghiem LD, Elimelech M. Osmotic dilution for sustainable greenwall irrigation by liquid fertilizer: Performance and implications. Journal of Membrane Science. 2015;494:32–8. DOI:10.1016/j.memsci.2015.07.026.

[82] Li S, Kim Y, Phuntsho S, Chekli L, Shon HK, Leiknes TO, Ghaffour N. Methane production in an anaerobic osmotic membrane bioreactor using forward osmosis: Effect of reverse salt flux. Bioresource Technology. 2017;239: 285–93. DOI:10.1016/j.biortech.2017.05.044.

[83] Lutchmiah K, Lauber L, Roest K, Harmsen DJH, Post JW, Rietveld LC, van Lier JB, Cornelissen ER. Zwitterions as alternative draw solutions in forward osmosis for application in wastewater reclamation. Journal of Membrane Science. 2014;460:82–90. DOI:10.1016/j.memsci.2014.02.032.

[84] Capodaglio AG, Olsson G. Energy issues in sustainable urban wastewater management: Use, demand reduction and recovery in the urban water cycle. Sustainability. 2020;12. DOI:10.3390/su12010266.

[85] Mamais D, Noutsopoulos C, Dimopoulou A, Stasinakis A, Lekkas TD. Wastewater treatment process impact on energy savings and greenhouse gas emissions. Water Science and Technology. 2015;71:303–8. DOI:10.2166/wst.2014.521.

[86] McCarty PL, Bae J, Kim J. Domestic wastewater treatment as a net energy producer-can this be achieved? Environmental Science and Technology. 2011;45:7100–6. DOI:10.1021/es2014264.

[87] Wan J, Gu J, Zhao Q, Liu Y. COD capture: A feasible option towards energy self-sufficient domestic wastewater treatment. Scientific Reports. 2016;6:1–9. DOI:10.1038/srep25054.

[88] Vicentin R, Fdz-Polanco F, Fdz-Polanco M. Energy integration in wastewater treatment plants by anaerobic digestion of urban waste: A process design and simulation study. International Journal of Chemical Engineering. 2019;2019. DOI:10.1155/2019/2621048.

[89] Lei Z, Yang S, Li Y you, Wen W, Wang XC, Chen R. Application of anaerobic membrane bioreactors to municipal wastewater treatment at ambient temperature: A review of achievements, challenges, and perspectives. Bioresource Technology. 2018;267:756–68. DOI:10.1016/j.biortech.2018.07.050.

[90] Lefebvre O, Quentin S, Torrijos M, Godon JJ, Delgenès JP, Moletta R. Impact of increasing NaCl concentrations on the performance and community composition of two anaerobic reactors. Applied Microbiology and Biotechnology. 2007; 75:61–9. DOI:10.1007/s00253-006-0799-2.

[91] Ansari AJ, Hai FI, Price WE, Ngo HH, Guo W, Nghiem LD. Assessing the integration of forward osmosis and anaerobic digestion for simultaneous wastewater treatment and resource

recovery. Bioresource Technology. 2018;260:221–6. DOI:10.1016/j.biortech.2018.03.120.

[92] Aslan S, Şekerdağ N. Salt inhibition on anaerobic treatment of high salinity wastewater by upflow anaerobic sludge blanket (UASB) reactor. Desalination and Water Treatment. 2016;57:12998–3004. DOI:10.1080/19443994.2015.1059369.

[93] Gao Y, Fang Z, Liang P, Zhang X, Qiu Y, Kimura K, Huang X. Anaerobic digestion performance of concentrated municipal sewage by forward osmosis membrane: Focus on the impact of salt and ammonia nitrogen. Bioresource Technology. 2019;276:204–10. DOI: 10.1016/j.biortech.2019.01.016.

Chapter 5

Recent Advances in Measuring and Controlling Biofouling of Seawater Reverse Osmosis SWRO: A Review

Rana H. Idais, Azzam A. Abuhabib and Sofiah Hamzah

Abstract

This study presents recent literature that sheds light on the SWRO membrane biofouling, Inventory of causes, consequences, measurement, and possible solutions. In particular, biofouling of SWRO is considered as one of the challenges faced by seawater desalination today. For seawater desalination, mitigating membrane biofouling is essentially required and yet to be overcome. Specific shortcomings and prospective solutions are reviewed towards understanding the biofouling mechanism, pretreatment impacts, level of assimilable nutrients, and real-time monitoring. Accordingly, this review aims to address recent advances in biological fouling measurements and control to better understand biofouling and the best ways of dealing with such a challenging issue.

Keywords: biofouling, membrane, pretreatment, reverse osmosis, seawater

1. Introduction

Beyond doubts, Seawater desalination is commonly considered as a significant method towards producing and supplying potable water across the globe, especially in areas like the Middle East and North Africa (MENA) region characterized by a dry climate, low precipitation, and insufficiency of surface water. Despite the availability of various desalination technologies, membrane technology presented by Reverse Osmosis (RO) witnessed significant growth dominating about 60% of the desalination industry worldwide. The newly developed RO membranes characterized by high rejection and high flow membranes were allowed to operate at high pressures (up to 80–90 bar), thereby making conversions to 55–60% economically feasible. Such advancements towards simplifying RO processes from a two-stage treatment change to a single-stage array which in turn reduced capital and operational costs [1–3].

The high demand and global climate change have contributed to water scarcity in a significant way. As such, 71% of the world's population live under conditions of moderate to extreme water shortage, and about 66% (4.0 billion people) live in severe water deficiency. This well-felt scarcity could be a binding limitation on the socio-economic development of many countries according to Goal #6 [SDG6] of the Sustainable Development Goals cover all aspects of managing water for fair access, sustainability, and environmental protection. Having said so, seawater desalination is reliably seen compared to other sources, especially with the

IntechOpen

long-term satisfaction tends to be achieved fully or partially for the demand in areas around the globe where water scarcity is felt, such as Australia, Southern Carolina, the Middle East, and Northern Africa [4].

Seawater desalination technology by RO is proven to be an extreme energy-efficient compared to other conventional thermal distillation methods and therefore is economically feasible. Membrane technologies application in the desalination industry has witnessed some rapid development and growth over the past 20 years. However, SWRO membranes are highly sensitive to the feedwater characteristics and to the concentration of certain organic compounds that potentially lead to membrane fouling phenomena [1, 5–7].

2. Desalination pretreatment

Pretreatment is crucial to SWRO, as it influences membrane efficiency and life expectancy by fouling reduction. Practically, it is essential to enhance the raw water quality before passing through RO vessels to promote high and effective performance. Yet, Membrane fouling and scaling remain challenging even though the perfect design and operating conditions can be significantly helpful. Both source and quality of feed water influence the pretreatment choice towards better fouling control. Technically, the silt density index (SDI) and turbidity are the two main parameters determining pretreatment performance [8–10]. In addition, microbial foulants characterization can be found in [11]. Pretreatment techniques are designed to eliminate the microbial loads on high-pressure membranes but may scavenge nutrients and potentially provide a suitable environment for microbial growth. A comparison of the bacterial community composition can, therefore, answer whether pretreatment compartments serve as inoculum for high-pressure membranes. Physical and chemical water treatment processing feed water in desalination industry is referred to usually as pretreatment, as shown in **Figure 1**, usually proceed by a series of methods: coagulation and flocculation, followed by granular media filtration (e.g., anthracite coal, silica sand, or garnet) and cartridge filtration. Biocides such as chlorine and Peracetic acid, in addition to ozone or ultraviolet (UV), can be used when biofouling is a problem [2, 11].

Furthermore, membrane biofouling cannot be removed by conventional pretreatment methods such as coagulation, flocculation, ultrafiltration, and cartridge filters (CF), as they are unable to remove all passing microorganisms tending to colonize the membrane. Sand filtration combined with chemical disinfection is more efficient in reducing microbial contaminants, including viruses, to acceptable levels meeting drinking water standards. Technically, the pressure-driven process presented by membrane filtration can provide high-quality drinking water. However, it could be faced with vital challenges including system demanding, relatively high cost, clogging, scale formation, and biofouling. Moreover, membranes have a limited lifetime regardless of how good they are and so they may not reject all pollutants of concern for drinking water after a certain time of operation even if microorganisms are successfully removed. One consideration in large-scale applications is that membrane filtration systems produce considerable amounts of more concentrated wastewater per unit of purified water. Having said so, membrane selection must take into consideration the nature of the contaminants associated or extracted [10, 12, 13].

Microbial colonization at the membrane surface is traditionally overcome by overall applying disinfection to the feed water. Ideally, any disinfectant should not be expensive or hazardous. However, it must be highly toxic to microbes with zero

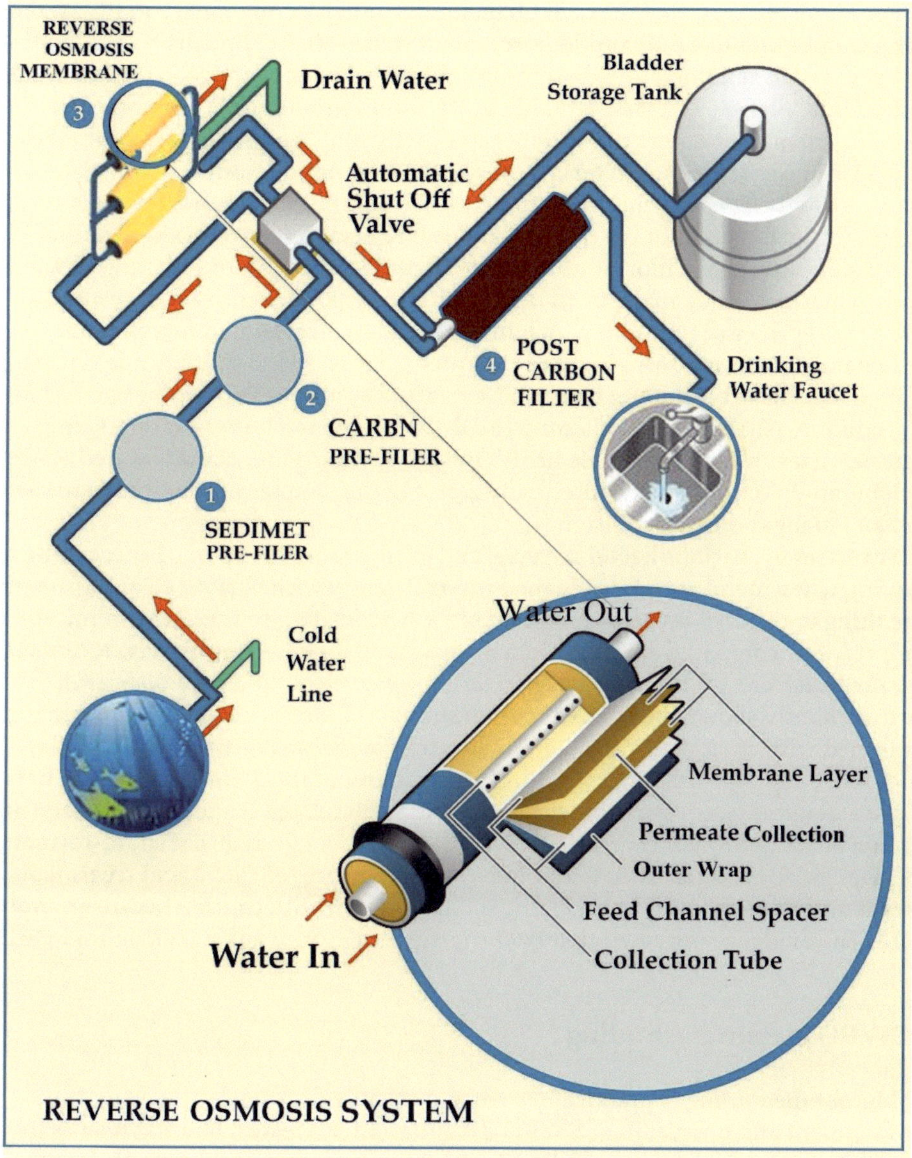

Figure 1.
Stages of an RO membrane system.

effect on the desalination plant productivity. Traditionally, there are many disinfection processes applied in water treatment including but not limited to chlorination, ozonation, and UV radiation. Although ozonation is found to be an effective disinfection technique using oxidative effects in removing microorganisms, it is a bit costly and unstable in addition to producing carcinogenic bromates as by-products in the treated water occasionally. Chlorine on the other hand is the most commonly used disinfectant characterized by easiness use and low cost. During the chlorination process, the biomass on the RO membrane is effectively destroyed. Besides and due to molecular analysis, some bacterial groups appear to tolerate this biocide. Well-known bacterial classes potentially resisting chlorination, such as Bacillus and Clostridia due to their ability to sporulate, are very much identified on fouled membranes [2, 14].

For many reasons, biofouling is challengingly difficult to manage in RO systems. Some membranes like polyamide-based membranes tend to be sensitive to oxidizing agents such as chlorine leading to significant limitations for such use. Generally, Commercial plants are not observably in sterile environments. Therefore, any microorganism that enters the system can rapidly multiply. Surprisingly, it takes only 30 minutes for some bacteria to duplicate their population, showing exponential growth. The free chlorine presented during the chlorination process may potentially lead to membrane damage and salt rejection deterioration. Another downside of applying chlorine as a disinfectant is its capacity of breaking down the organic and humic material to Assimilable organic carbon (AOC), resulting in the rapid growth of biofilm which in-turn leads to accelerated incremental of feed channel pressure drop. In some treatment plants, Mono-chloramine is usually applied to achieve biofouling control. Nevertheless, mono-chloramine can be used to produce N-Nitrosodimethylamine (NDMA), which is a human carcinogenic material that can result in public health issues. Furthermore, contaminated water with mono-chloramine may potentially result in the damage of membrane in the iron and manganese presentation [8, 12, 15].

Surprisingly, various bacterial types and groups were found to be succeeding and thriving when membranes are cleaned intermittently with various cleaning agents. One thing to think of is the inclusion of citric acid leading to several community compositions compared to when chlorine was used alone. *Acinetobacter, Ralstonia, Comamonadaceae,* and *Diaphorobacter, Stenotrophomonas*, and *Enterobacteriaceae* are dominantly shown on cleaned membranes via chlorination. When chlorination combined with citric acid cleaning Silicibacter, *Rhodobacteraceae, Pseudomonas, Pedobacter*, and *Janthinobacterium,* they became abundant. This is confirmed based on physiological features assigned to taxonomically related bacteria and Adenosine Triphosphate (ATP) concentrations. It is, therefore, suggested that spore-formers, Gram-positive bacteria, and Acidophiles are better resisted citric acid treatment. These suggestions should be taken into consideration with caution because simply, there is no evidence provided that bacteria are recalcitrant against citric acid [16].

3. SWRO membrane fouling

Membrane fouling is practically seen as a chronic drawback hindering the development and operation of SWRO desalination processes. Fouling results in overall membrane performance deterioration with operational pressure drop and more frequent cleaning leading to operational costs increase and eventually full loss of membrane. From hydrodynamics perspectives, fouling development mainly in space-filled channels of the membrane is influenced by water quality, operational conditions, and spacer and membrane design. Technically, membrane fouling issues vary from organic and inorganic fouling to colloidal and biofouling contributing to increase cost of operation as well as affecting the quality of water produced. Amongst, biofouling seems to be way too complicated and hard to be controlled due to the excessively increase of biofilm formation on the surface of membrane surface, consequently leading to deteriorated performance. Additionally, the capability of lived bacteria inside biofilms in terms of high tolerance to antibiotics and other antimicrobials than planktonic cultures is problematic. As such, various techniques including pretreatment, membrane surface alteration or modification, disinfecting of feed water via chlorination, and cleaning are developed to overcome and/or control biofouling simply by treating biofilm formation on membrane surfaces [2, 17, 18].

4. Biofouling

4.1 Definition

Presently, several foulants considered or categorized as microbial ones including various microorganisms and organic compounds, known also to be aquatic, such as polysaccharides, proteins, and lipids, called extracellular polymeric substances (EPS). Identically, the biofouling process involves in adhesion of organisms that are aquatic along with their metabolic products presented on membrane surface or feed spacers. As shown in **Figure 2**, strong biofilm growth can be observed and found on the feed spacer strands. More than 45% of all membrane fouling is biofouling originated mainly by unicellular or multicellular microorganisms and therefore seen as one of the major issues of concern to reverse osmosis membrane filtration processes. Although membrane biofilm majority is formed by bacteria, other organisms such as fungi, algae, and protozoa may potentially be attracted by the membrane surface and add up to the formation of biofilm in a significant manner. Various studies confirm that *Pseudomonas, Bacillus, Arthrobacter, Corynebacterium, Flavobacterium,* and *Aeromonas* are the most predominant bacteria identified in fouled RO membranes [9, 11, 12, 19–22].

Membrane biofouling takes place gradually in sequential steps, as shown in **Figure 3**. Firstly, the microbial cells get to membrane surface attachment, causing the forming of the biofilm as layer, involving communities of different microorganisms' types (e.g., bacteria, algae, protozoa, and fungi). Initially and acting as mediator for the attachment of microbial substances, electrokinetic and hydrophobic interaction, the growth and multiplication of cell usually follows at the expenses of nutrients being soluble in water feed or membrane surface adsorbing organics. The roughness and charge of the membrane surface are considered as key factors contributing to the enhancement of the microorganisms attached to the membrane surface [9].

Several environmental factors raising bacterial growth such as nutrients amount and types which strongly affect the microbial composition and biofilms density. Also, membrane characteristics such as type, roughness, charge, and hydrophobic/hydrophilic characters very much influence the biofouling microbial film establishment. Producing RO membranes highly resistant to biofouling as well as other fouling types remains challenging. Typically and from operational point of view, biofouling poses itself as a challenge, especially for saline waters having natural organics at high levels. Seasonally, biofouling tends to be problematic during extreme algal blooms or in time of having accident entrance to the open intake of the plant in rainy season with highly organic water [2, 11].

Figure 2.
Biofouling in RO sample (left: top view, right: cross-section) [12].

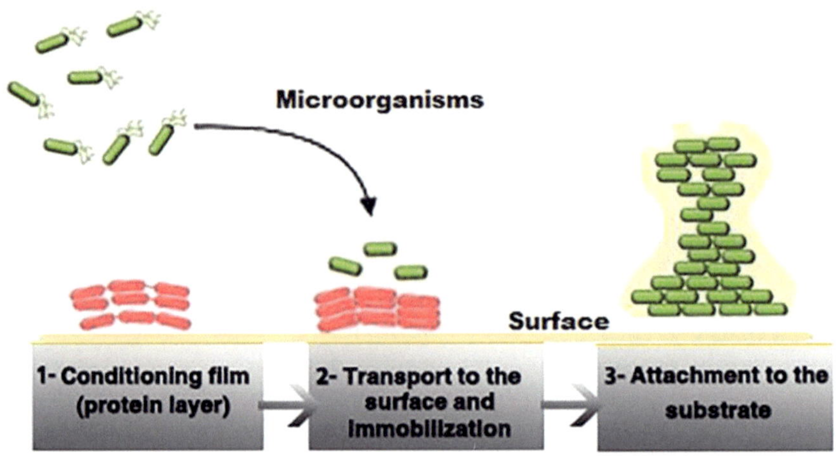

Figure 3.
Steps of the biofilm [20].

Commonly, biofouling attributes to the increased probability of bacteria producing polysaccharides and natural adhesives. It occurs at all open-ocean desalination plants such as the Jeddah SWRO desalination station in KSA. Mature biofilms exhibit anti-bactericidal properties and are also resistant to detachment. Biofilm formation results in biofouling when exacerbated in desalination systems by water production efficiency deterioration of membrane degradation, leading to a significant increase in operational costs associated with cleaning regiment and shortened membrane lifespan [23, 24].

4.2 Factors affecting biofouling

Generally, the saline feed water biofouling potential is influenced by several interrelated factors including microorganisms' concentration; content of readily biodegradable compounds; nutrients concentration and composition in the source water; temperature; the salinity of the feed water as well as operating parameters such as cross-flow velocity [11, 25, 26].

The study of [25] elucidated Algal organic matter (AOM) impact on biofouling affecting various membranes modules (capillary and spiral wound) by algal blooms. They found that measuring Adhesion force illustrate that AOM has the propensities towards adhering to a membrane surface and would need massive force to be removed from the membrane. Also, the seawater capacity supporting bacterial growth illustrated a correlated positive linear with AOM concentration levels in the water. It was linked to the tending of AOM, especially, transparent exopolymer particles (TEP), to nutrients concentration absorption from the feed water feeding attached bacteria. Also, fastened experiments of biofouling made with spiral wound and capillary membranes evidently show that when biodegradable nutrients presented in the feed water unlimitedly, a high level of AOM concentration in water feeds or as membrane attachment may significantly speedup biofouling. Further observation is that lower biofouling rates occurred once membranes are exposed to feed spikes with AOM or nutrients [25].

4.3 Microbial communities in RO systems

The bacteria can tolerate a wide range of pH (0.5–13) and temperature (−12–110 °C) while being able to colonize on all membrane surfaces in RO plants under different conditions. Various studies were carried out to investigate frequently observed microorganisms on membranes in RO plants. As concluded by [28] some bacterial groups are presented with some potential finger print significantly related to biofouling. Their study mostly opened some future window towards focusing on having already-cleaned membranes treated prior to installation. Also, paying more attention to primarily target troubling colonizers, or developing pre-treatment designs considering biofouling measures through the bacterial load minimization attempting to access membrane unit feed. While, a pilot-scale study of [29] was implemented to compare bacterial populations (membrane biofilm) in seawater, CF, and from Carlsbad plant at California, USA.

Observably, population of biofilm for seawater and membrane tend to have some similarity, but the CF harbored other biofilm community type. It was a relatively firm study because it concluded the findings of different communities of biofilm in five fouled SWRO membranes than those of other found around the globe. Apparently, such unique occurrence was due to differences observed in operational conditions and sampling across the year. Various mutually and dominantly existence of bacterial group could be observed in all samples. As such, strong suggestion was made about certain group being conformed to the membrane surfaces growing under chemolithoheterotrophic conditions oligotrophically [27–29].

Similarly, [30] results found that members of the *Ruegeria, Pseudoruegeria, Parvularcula, Legionella,* and *Shigella* were the only bacterial groups shared between the CF and RO membranes. *Phaeobacter*, *Leisingera, Kangiella,* and *Bacillales* are abundant in the CF, while *Haliangium* and *Limnobacter* are abundant on the RO membrane. The presence of bacteria belonging to taxa harboring facultative and obligate chemolithotrophs, such as *Geobacter, Desulfuromusa*, and *Thioalkalivibrio,* on the CF potentially indicate the effective removal made by the pre-treatment compartments for certain nutritional compounds, such as ferrous iron or sulfurConsequently, published studies do not uniformly present the composition of the bacterial community at the same taxonomic level, hardening the comparison of bacterial diversity. For instance, a review of [14] compared 33 studies investigating bacterial communities on fouled high-pressure membranes. They classified the identified bacteria at the order level. A total of 35 bacterial orders from those fouled high-pressure membranes have been recorded. These orders were used as a benchmark to compare the microbial diversity of feed water, pre-treatment compartments, and fouled membranes, and to detect the role of specific selection pressures on the microbial composition [14, 30].

A review of [14] found that the most commonly detected bacteria on fouled membranes are *Burkholderiales, Pseudomonadales, Rhizobiales*, and *Sphingomonadales,* and *Xanthomonadales.* Whereas *Burkholderiales* and *Xanthomonadales* have not been identified in earlier studies, but studies of next-generation sequencing (NGS) have frequently identified these orders of bacteria on fouled membranes. Due to its ability to study bacterial community compositions in a culture-independent and high-performance way [31]. In [32] they compared the bacterial diversity of the surface water and the membrane population. They concluded that the biofilm actively produced on the membrane surface, rather than being a concentration effect of bacteria. In general, the composition of the bacterial population on the membrane varies from the feed water because only a fraction of the bacterial feed water diversity accumulates at the membrane surface, indicating that the membrane surface provides bacterial selection pressures. However, [33] found that the bacterial composition of a mature fouling layer was similar to the feed water composition [14, 31–33].

In the experiment of [34], a lab-bench cross-flow RO system was used to explore the impact of chlorine disinfection on reverse osmosis membrane biofouling.

No significant distinctively chlorine-resistant bacteria were detected in the sample without chlorine dosage and with 1 mg-Cl_2/L chlorine dosage. However, in the samples with 5 and 15 mg-Cl_2/L chlorine, kinds of significantly distinctive chlorine-resistant bacteria were presented included *Methylobacterium, Pseudomonas, Sphingomonas*, and *Acinetobacter.* These results indicated the significant selection effect of chlorine on the chlorine-resistant bacteria. Results of [35] found *Proteobacteria*, *Bacteroidetes*, *Firmicutes,* and *Planctomycetes* are the most abundant phyla with the application of high throughput sequencing. Microbial community succession was revealed during biofilm formation, in which *Proteobacteria*, *Planctomycetes,* and *Bacteroidetes* played significant roles [34, 35].

The research of [28] analyzed the biochemical properties by selecting a good-model bacteria include *Paracoccus, Burkholderia, Pseudomonas, Acinetobacter, Pseudoalteromonas, Cytophaga, Microbacterium, Bacillus, Marinomonas, Rhodococcus, Exiguobacterium,* and *Staphylococcus* which may influence its ability in terms of forming insurgent biofilms cumulatively at membrane surfaces. In this study, bacteria was isolated across stages of all plant. Predominant organisms were detected and seem to be significantly involved in biofouling as well as including almost all isolated cultures by culturing and next-generation sequencing (NGS) through applying 16S rRNA meta-barcoding. Researchers have also found that as biofilm community influenced by bacterial community on seawater reverse osmosis membranes, it is compulsory to have customized/designed controlling measures targeting the invading microbial elements related to the plant's geographical spot [28].

4.4 Biofouling potential indicators and measurements

A biofilm has a high content of water and organic matter (70–95%), high numbers of colony-forming units and cells, high contents of carbohydrates and proteins, high content of ATP, and low content of inorganic matter. Indicating biofouling potential can be proposed by multiple parameters as ATP, AOC, and biodegradable dissolved organic carbon (BDOC). Generally, the previously mentioned parameters are generally applicable for fresh waters and yet to be extended to be applied for desalination plants [25, 36]. Meanwhile, the study of [37] suggested some testing sets to allow for the determination of the water samples capacity of microbial support. In addition to using fluorescence intensity microplate analysis to determine biofouling potential on RO membranes [35, 37, 38].

Measurement of RO feed water biofouling tendency is not an easy task. To do so, several in-practice parameters are indicatively considered like: Silt density index SDI, turbidity, and total suspended solids (TSS). Having said so, biologically-based data is yet to be obtainable supporting such measurements. The RO feed water microbial support capacity (MSC) is practically determined by factors associated with the algal activity, such as TOC, the ratio of TOC:TN: TP, the increase in RO train DP, Chlorophyll a, TEP, bacterial activity (e.g., ATP), total bacterial count, microscopic observation, and nutrients concentration (Total N, Total P). Biological-based factors such as AOC and BDOC are used in waters with no salinity. Also, many consistent monitoring systems like monitoring of biofilm and the MFS had been developed to determine formation rate of biofilm. These monitoring systems cannot predict the feed water potentiality for biofouling but simulate overall plant operation [11, 19].

The concentration of TOC is widely applied to indicate the potentiality of saline water to biofouling whereas the rate of DP increase is indicatively used for the rate of biofouling. From operational point of view, potentiality to biofouling tends to be significantly increased when TOC concentration raises to 2 mg L^{-1}.

Practically, the weekly measurements made for ratio of TOC:TN: TP to indicate biofouling increasing. Consequently, ratios above 20% of 1:1:1 indicates an elevation requires bacterial EPS generation leading to have the bacteria encouraged to cause membrane fouling [11, 39].

The concentration of Chlorophyll *a* for the feed water can be indicatively seen as a sign related to the green pigmentation algae content in the water [11]. The total count of algae is potentially determined via online methods or through lab experiments. Technically, there are three KPIs (key-performance-indicators) determining the algal high content impact in feed water including: efficiency of solid removals deterioration at pre-treatment stage due to filtration overload, fouling acceleration of CF, and finally RO train productivity deterioration [11].

4.5 Membrane biofouling impact

In SWRO systems, biofouling has many adverse effects, as increases in differential driving pressure and feed channel pressure drop. These are required to maintain the same production rate due to biofilm resistance. In addition to increased energy consumption associated with high pressure to achieve the biofilm resistance and flux decline. Biofouling eventually leads to the biodegradation of cellulose acetate membranes caused by acidic by-products concentrated at the membrane surface. Also, it leads to reducing the active membrane area, and therefore decreased flux of permeate due to the formation of a low-permeability biofilm on the membrane surface. Other main consequences of biofouling decreased membrane permeability, increased the frequency of chemical cleaning, and the possible increase in replacement frequency of membrane [9, 19, 24, 40].

Research conducted by [41] investigated the biofouling effect on the sequentially declining in reverse osmosis membranes in terms of membrane operational parameters like membrane permeability, pressure drop in feed, salt rejection. Also, the consumption of temporal organic carbon (DOC) is being measured. It could be illustrated that all indicators were influenced by biofouling formation. Observed increase in the pressure drop in the feed channel (FCP) affected permeability and decline salt rejection, consequently leading to prove the FCP sensitivity to biofouling. Besides, [35] found that biofouling can accelerate the formation of scaling, and the mixed foulants can block the membrane pores, leading to a significant flux drop [35, 41]. In brief, biofouling has a potential effect on the following: differential driving pressure, feed channel pressure, energy consumption, the flux of permeate, membrane area, membrane permeability, the frequency of chemical cleaning, and salt rejection.

4.6 Biofouling alleviation and control

The control of biofilm formation is a complicated and controversial process involving the reduction of microorganisms within the RO water, monitoring strategies, and controlling factors such as nutrient concentrations and physicochemical interactions between microorganisms and membrane surface. Gulf Sea at the Saudi Arabia is known to be having biofouling major challenge uneasy to be controlled. It still the main challenge in membrane filtration installations. Curative or preventive measures are not always efficient. Flocculants provide a suitable habitat for microbial growth, whereas conditioning agents are potential sources of microorganisms and nutrients for the biofilm. Another source of microbial contamination is the piping, storage tanks, and treatment systems before RO, such as ion exchangers and active carbon filters. Biofilm can grow in very low-nutrient habitats with TOC levels as low as 5–100 μg/L. In practice, several methods for biofouling control have been investigated, such as the application of the pretreatment before SWRO to remove bacteria and biodegradable organic matter, dosing of biocides, and limiting essential nutrients such as carbon and phosphate [9, 40, 42–44].

Membrane cleaning as a method of biofouling control typically done when there a significant decrease in differential pressure drop or permeability. Principally, cleaning process involves removing and/or destroying of the biomass accumulating on membrane surface to reserve membrane permeability. Cleaning process can be applied physically or chemically. Physical cleaning was usually performed before chemical cleaning, involving flushing of air and water. It requires applying pressure mechanically, attributing to the removal of all non-adhesive fouling-based. Membrane manufacturers suggest different chemical agents' forms for cleaning purposes (e.g. alkaline, acids, biocides, enzymes, and detergents). Such process is efficiently eliminating or deactivating non-accumulating microorganisms. Therefore, the residual inactive biomass can be consumed as food by survived bacteria leading to bacteria regrowth acceleration. Base/acid cleaning removes organic foulants on membranes and destroys the microbial cell walls. Metal chelating agents and surfactants were used to disintegrate EPS layers by removal of divalent cations and solubilization of macromolecules, respectively. The efficiency of cleaning agents to remove biofouling is limited because the EPS layer is recalcitrant against cleaning agents. Improvement of cleaning efficiency difficult, particularly for aged biofilms. Membrane cleaning frequently removes only part of the fouling layer and cleaned membranes, therefore, provide a suitable environment for swift microbial colonization. Thus, cleaning processes (physically and chemically based) may partially result in biofouling reduction on the short run without sustainably controlling biofouling on the long run [8, 45–47].

Control of bacterial growth by chemical disinfectants depends on many factors, such as chemical concentration, its mode of action, contact time, the density of organisms, and TSS of feed water. These factors make it extremely difficult to attain absolute disinfection. Besides, chemical disinfectants like chlorine and its derivatives may be hazardous to health. Chlorine is known to oxidize and degrade the humic substances in seawater, thus, resulting in smaller molecules, which are AOC. The AOC is a good nutrient source for marine bacteria, and under such status could also lead to rapid biofilm formation in SWRO plants. Chlorination may foster the formation of trihalomethanes and other chlorinated by-products, which are carcinogenic [48].

Many researchers have concluded that biofouling is inevitable and tend to be difficult to prevent with having the focus shifted towards control strategies aiming to achieve: biofilm formation delay, biofilm accumulation impact reduction or delay on performance, and finally removing biofilm via advanced strategies of cleaning. For many reasons, biofouling control tends to be challenging. As such, various methods were developed towards treating biofilm formation on membrane surface and/or mitigating biofouling effect in general. Instantly, some strategies were applied including: membrane flushing or cleaning, application of chemical additives to target bacterial cell or extra-cellular matrices, membrane surface modification, limiting nutrient content, and the quenching of quorum. All previously mentioned methods have limitations and may result in unwanted membrane degradation [14, 18, 21, 49, 50]. As part of chemical treatments with biocides in addition to anti-microbes were applied mutually as part of industry practices. Chemically-based cleaning are known to be affecting exclusively the topper biofilm layers by colonizers. The effect of nutrient levels and possible manners to control membrane biofouling poses another potential solution for many membrane installations and should be further investigated. Biofouling impact on membrane efficiency is potentially minimized through a combination of strategies involving early identification, preventive cleaning, substrate limitation for delaying biofouling built-up, and cleaning procedures optimization towards effective biofilm removal [14, 41, 50].

Based on the current knowledge, membrane surface modifications tend to be incompatible for control biofilm formation in full-scale membrane operations because of the drag force that transfers bacteria and nutrients to the membrane surface. As various components are moved to the membrane surface by the drag force, they are easily covered, and membrane surface modifications are rendered less efficient. By applying comprehensive pretreatment, therefore, biofouling can be limited but not eliminated. Practically, membrane biofouling prevention tends to be fully or partially achievable by better pretreatment in new desalination systems. Yet, it might be essential to have old, insurgent biofilms and prolonged membrane operating plants dispersed sufficiently. Most existing techniques in efficiently use an enormous spectrum of biocides and chemicals attacking bacteria to maturely disperse biofilms [14, 26, 28].

Practices presented as clean-in-place (CIP) tend to be less efficient and that successful. This is related probably to various reasons including: wrong selection of chemical, improper pH control, temperature, time of contact, unsuitable recirculating flow rates, and partial biomass removing. The repetitive biocides usage potentially lead to bacterial resistance inducing via bacterial cell modification on membrane surface, permeability deterioration of biocide, and biocides degradation by enzymes development, or gaining more resistance for biocide genes [28].

Strategies for Biofilm control applying enzymes towards degrading of EPS matrix including glycosidases, proteases, and deoxyribonucleases. However, these enzymes targeting specific strains, and their sufficiency in complex multi-species biofilms is yet to be established. Also, enzymes are costly and may not be so practical when applied for membrane treatment or flushing. On the other hand, a bad need for more efficient and cost-effective methods to eliminate biofilms and alleviate biofouling in SWRO processes do exist. As such, it is highly recommended to conduct researches investigating novel chemical cleaning agents which may positively contribute to overcome or mitigate biofouling [26, 28].

A study of [18] investigated the *Pseudomonas* quinolone signal (PQS) pathway role in biofouling control in reverse osmosis membranes. They inoculated *Pseudomonas aeruginosa* inside water feed as a sort of biofouling simulation. Conversely, a novel-based method on quorum sensing (QS) biochemically, has triggered considerable interest in controlling biofouling. Several advantages could be concluded, as it is characterized by high efficiency low operational pressure, practically contribute to hindering development capacity of the bacteria. QS Identified as cell-to-cell signals whereby microorganisms applied it for the sake of cell density sensing; reaching to a critical threshold level in terms of signals will trigger responsive sets of genes. Many researchers have found that the interference with such cell density-dependent communication technique formulate a biofilm potential controlling strategy.

The application of bacteriophage in synergetic way combined with some other traditional methods, such as cleaning proven to be mitigating *P. aeruginosa* biofouling-based sufficiently. Some alternative options are presented by bacteriophages. *Pseudoalteromonas*, for example, presented in high amount on a marine-based biofilm layer is potentially isolatable and known to be having some lytic footprint, highly efficient. Lastly, it might be of interest to explore the bacteriophage treatment effectiveness in biofilm formation prevention instead of having the structure of biofilm removed. To this regard, bacteriophage activation may be limited by low cell density. Another bacterial hosts might be targeted by taxonomical families performing a more sufficient approach towards maximizing infection impact on the way to achieve biofouling mitigation [51].

In [51] research, the isolation of lytic bacteriophages was used to hinder *P. aeruginosa* growth in planktonic-based mode and varied pH, salinity level, and temperature. Accordingly, bacteriophages were found to be optimally infective with 10-times infection multiplication under salinity mode. It illustrated that the lytic has reasonable abilities over experimental testing temperatures (25, 30, 37, and 45 °C) and pH range of 6–9. When exposed to bacteriophages, Planktonic *P. aeruginosa* found to be significantly exhibiting a longer lag mode and low rates of specific growth, taking into account the application of bacteriophages to P. made in subsequent manner. The biofilm presented by *aeruginosa*-enriched was determinant to lowering the relative amount of Pseudomonas-related taxa from 0.17 to 5.58% in controlling to 0.01–0.61% in processed communities of microbes. The findings illustrated the potential application of bacteriophages as a biocidal agent to achieve the mitigation of unwanted *P. aeruginosa* associated with issues in seawater-based applications [51].

In [26] study, biofilm amount and characterization were analyzed concerning membrane performance applying acid/base cleaning. Generally, cell and tissue of the bacteria deactivate chemical agents used in cleaning process to remove mainly the biomass related to biofouling. Chemical-based reactions like dispersing, chelating, solubilization, suspension, peptisation, sequestration, and hydrolyzing are observed during cleaning process. Cleaning by Alkaline-based solutions like Sodium hydroxide was also applied in this study for three types of biofilm to explore biofilm removal efficiency as well as illustrating EPS matrix role. They concluded that with minimum biomass amount at low substrate concentration cleaning was not as efficient as with high substrate concentration, with same observed phenomena for membrane performance restore [26].

While [43] describes the biofouling monitoring technology of the "Megaton Water System" project and verifies the technology in the pilot and real plants in Al Jubail, Saudi Arabia. Biofouling monitoring technology refers to the community of bacteria composition change by chemical usage of the Membrane Biofilm Formation Rate (MBFR) was applied to this project was a positive indication of a reliably system design and operation. Such monitoring technology could be applied to achieve plant operational and reliability improvement throughout the overcome of biofouling issue. It could also assist in environmental impact reduction and lower plant production costs through chemical-free injection [43].

According to [52] study, they developed a simple method where a combination of bubbling and cleaning-based on frequent addition of hydrogen peroxide (H_2O_2) at lower concentration level at feed water. The same approach was also explored with the use of CuO or PP spacers. The dosage of 0.3% (w/w) H_2O_2 being applied periodically at 12 h intervals resulting in having no increase in FCP in the tested system, also an indication referring to the tangible biofouling lacking with intermittent H_2O_2 dosing. For tested fouled membranes fouled over a period of eleven days, a single dose of 0.3% (w/w) H_2O_2 applied and successfully eliminated almost all spacers and membranes accumulated biofilm in few minutes demonstratively by a FCP of 69% (CuO spacer) and 54% (PP spacer). The biofouling reduction was primarily due to the high shear created by the generated oxygen bubbles in the system, combined with the disinfection effect of H2O2. The reasonably low cost of $0.009/m^3 from intermittent H_2O_2 dosage was not more than 0.8% of overall assumed cost and 6.5% out of pre-treatment cost, allowing for economical accepted approach to overcome biofouling [52].

It seems that dechlorination water, activated carbon, cartridge filtration, UV irradiation, ozone treatment, hydrogen peroxide, detergents, alkaline, sodium bisulfite, and hot water sanitization are effective techniques and limitations to control biofouling.

5. Conclusion

Biofouling in SWRO membranes continues to be problematic for operation and maintenance quality. It plays an essential role in the fouling of the membrane parts in full-scale and pilot-scale plants, and it's significant to reduce its occurrence by prediction and prevention. The study demonstrates the RO membrane biofouling mechanisms and the effective fouling control strategies within seawater desalination, where biofouling is a critical drawback. The study aims to evaluate microbial fouling (biofouling) to understand its effect on RO membrane performance. The study highlighted the composition of the microbial community and the functional potential of the RO membrane biofilm. In general, biofouling has affected all performance indicators. The selection of pretreatment seems to be a factor affecting the microbial community composition and functional potential. Analysis of the biofilm bacterial community has shown that seasonal changes in water quality influenced the biofouling bacteria.

The results showed that the accumulation of biofilms on membrane surfaces remains the key obstacle for high-pressure membrane filtration. For future research, it is significant to describe the cleaning agent and cleaning frequency. Also, measuring feedwater temperature, determine the location of the membrane element, and the sampling location at the membrane. These comprehensive researches will use to establish an integrated strategy to control biofouling. Biofouling control should concentrate on improving low fouling feed spacers, and the hydrodynamic conditions reduce the effect of biomass accumulation.

We conclude that to maintain plant productivity and membrane recovery it is necessary to increase the membrane cleaning frequency. In the CF and RO membrane, the microbial regrowth rate is a significant factor that impacts the biofouling rate. We recommend further searches of the strategy of balancing the nutrient levels as a solution for several membrane installations to control membrane biofouling. To measure biofouling, it needs for real tool, sensitive pressure drop data, and systematic methodology. Therefore further studies related to avoiding adverse biofouling processes will be valuable to investigate specific microbial members in more detail using biofilm monitoring and control strategies. Finally, additional SWRO research and development are critical for the efficiency of this growing industry.

Abbreviations

AHL	N-acyl-homoserine lactone
AOC	Assimilable organic carbon
ATP	Adenosine Triphosphate
BDOC	Biodegradable Dissolved Organic Carbon
CF	Cartridge Filters
CIP	Current Clean-In-Place
DOC	Temporal Organic Carbon
DP	Differential Pressure
EPS	Extracellular Polymeric Substances
FCP	Feed Channel Pressure
H_2O_2	Hydrogen Peroxide
MENA	Middle East and North Africa
MFS	Membrane-Fouling Simulator
$\mu g\ L^{-1}$	Micrograms Per Liter

mg L^{-1}	Milligrams Per Liter
MSC	Microbial Support Capacity
NDMA	Nitrosodimethylamine
PQS	*Pseudomonas* Quinolone Signal
PX	Pressure Exchangers
QS	Quorum Sensing
RO	Reverse Osmosis
SDI	Silt Density Index
SWRO	Seawater Reverse Osmosis
TEP	Transparent Exopolymer Particles
TOC	Total Organic Carbon
TSS	Total Suspended Solids
UF	Ultrafiltration
UV	Ultraviolet

Author details

Rana H. Idais[1], Azzam A. Abuhabib[1*] and Sofiah Hamzah[2]

1 Water Technology PhD Joint Programme between Islamic University of Gaza (IUG) and Al-Azhar University of Gaza (AUG), Gaza, Palestine

2 Faculty of Ocean Engineering, Technology and Informatics, Universiti Malaysia Terengganu, Terengganu, Malaysia

*Address all correspondence to: azz200@hotmail.com

IntechOpen

© 2021 The Author(s). Licensee IntechOpen. This chapter is distributed under the terms of the Creative Commons Attribution License (http://creativecommons.org/licenses/by/3.0), which permits unrestricted use, distribution, and reproduction in any medium, provided the original work is properly cited. CC BY

References

[1] Lattemann S. Development of an environmental impact assessment and decision support system for seawater desalination plants [thesis]. UNESCO-IHE: Delft University of Technology: the Netherlands; 2010.

[2] Al-Abri M, Al-Ghafri B, Bora T, Dobretsov S, Dutta J, Castelletto S, Rosa L, Boretti A. Chlorination disadvantages and alternative routes for biofouling control in reverse osmosis desalination. Clean Water. 2019 ; 2: 2. DOI:10.1038/s41545-018-0024-8

[3] Oussama N, Bouabdesselam H, Ghaffour N, Abdelkader L. Characterization of seawater reverse osmosis fouled membranes from large scale commercial desalination plant. Chemistry International. 2019; 5: 2: 158-167. https://doi.org/10.5281/zenodo.1568742

[4] World Bank. 2019. The Role of Desalination in an Increasingly Water-Scarce World. World Bank, Washington, DC. Available from: www.worldbank.org/water

[5] Dehwah A.H, Li S, Al-Mashharawi S, Winters H, Missimer T. M. Changes in feedwater organic matter concentrations based on intake type and pretreatment processes at SWRO facilities, Red Sea, Saudi Arabia. Desalination. 2015; 360: 19-27. DOI: 10.1016/j.desal.2015.01.008

[6] Dehwah A.H, Missimer T. M. Subsurface intake systems: Green choice for improving feed water quality at SWRO desalination plants, Jeddah, Saudi Arabia. Water Research. 2016; 88: 216-224. DOI:10.1016/j.watres.2015.10.011

[7] Al-Kaabi H, Mackey H.R. Environmental assessment of intake alternatives for seawater reverse osmosis in the Arabian Gulf. Journal of Environmental Management. 2019; 242: 22-30. DOI:10.1016/j.jenvman.2019.04.051

[8] Abushaban A, Salinas-Rodriguez S, Dhakal N, Schippers J, Kennedy M. Assessing pretreatment and seawater reverse osmosis performance using an ATP-based bacterial growth potential method. Desalination. 2019; 467: 210-218. https://doi.org/10.1016/j.desal.2019.06.001

[9] Ismail A, Khulbe K, Matsuura T. RO Membrane Characterization. Handbook of Reverse Osmosis. Elsevier; 2019. p 57-90. DOI:10.1016/b978-0-12-811468-1.00008-6.

[10] Badruzzaman M, Voutchkov N, Weinrich L, Jacangelo J. G. Selection of pretreatment technologies for seawater reverse osmosis plants: a review. Desalination. 2019; 449: 78-91. DOI:10.1016/j.desal.2018.10.006

[11] Voutchkov N. Membrane Foulants and Saline Water Pretreatment, Handbook of Pretreatment for Reverse Osmosis Desalination. Elsevier; 2017. p. 11-41. http://dx.doi.org/10.1016/B978-0-12-809953-7.00002-4

[12] Gassol G. M. Biofouling control in reverse osmosis membranes for water treatment [thesis]. Universitat Rovira I Virgili; 2017.

[13] Cahoon L. Water Purification: Treatment of Microbial Contamination. Handbook of Advances in Water Purification Techniques. Elsevier; 2019. p. 385-395. https://doi.org/10.1016/B978-0-12-814790-0.00015-6

[14] Vries H. J, Stams A.J.M, Plugge C.M. Biodiversity and ecology of microorganisms in high pressure membrane filtration systems. Water Research. 2020; 172: 115511. https://doi.org/10.1016/j.watres.2020.115511

[15] Sanawar H, Xiong Y, Alam A, Croué J.-P, Hong P.-Y. Chlorination or monochloramination: balancing the

regulated trihalomethane formation and microbial inactivation in marine aquaculture waters. Aquaculture. 2017: 480: 94-102. DOI: 10.1016/j.aquaculture.2017.08.014

[16] Khan M. T, Hong P, Nada N, Croue J. Does chlorination of seawater reverse osmosis membranes control biofouling. water research. 2015; 78: 84-97. http://dx.doi.org/10.1016/j.watres.2015.03.029

[17] Rachman R.M, Li Sh, Missimer Th. M. SWRO feed water quality improvement using subsurface intakes in Oman, Spain, Turks and Caicos Islands, and Saudi Arabia. Desalination. 2014; 351: 88-100. DOI:10.1016/j.desal.2014.07.032

[18] Li Ch, Liang J, Yang Y, Pu J, Hou L. Novel insights into the role of Pseudomonas quinolone signal in the control of reverse osmosis membrane biofouling. Journal of Membrane Science. 2018; 563: 505-512. https://doi.org/10.1016/j.memsci.2018.06.005

[19] Tabatabai S. Coagulation and ultrafiltration in seawater reverse osmosis pretreatment [thesis]. UNESCO-IHE: Delft University of Technology: the Netherlands; 2014. http://resolver.tudelft.nl/uuid:60173b3b-2d8a-4194-acf0-57ced0ee4968

[20] Pichardo-Romero D, Garcia-Arce Z, Zavala-Ramírez A, Castro-Muñoz R. Current Advances in Biofouling Mitigation in Membranes for Water Treatment: An Overview. Processes. 2020; 8: 182. DOI:10.3390/pr8020182

[21] Matin A, Khan Z, Zaidi S.M.J, Boyce M.C. Biofouling in reverse osmosis membranes for seawater desalination: phenomena and prevention.Desalination. 2011; 281: 1-16. DOI: 10.1016/j.desal.2011.06.063

[22] Hong P.Y, Moosa N, Mink J. Dynamics of microbial communities in an integrated ultrafiltration–reverse osmosis desalination pilot plant located at the Arabian Gulf. Desalination and Water Treatment. 2015; 1-14. DOI: 10.1080/19443994.2015.1083483

[23] Mohamed H. E, Al-Sharif Sh. F, Bamaga O. A, Albeirutty M. H. Potential biofoulants in open-ocean SWRO desalination station in Jeddah, KSA. Water Science & Technology: Water Supply. 2018. DOI: 10.2166/ws.2017.246

[24] Khan B.K, Fortunato L, Leiknes T. Early biofouling detection using fluorescence-based extracellular enzyme activity. Enzyme and Microbial Technology. 2018. https://doi.org/10.1016/j.enzmictec.2018.10.001

[25] Villacorte L. O, Ekowati Y, Calix-Ponce H. N, Kisielius V, Kleijn J. M, Vrouwenvelder J. S, Kennedy M. D. Biofouling in capillary and spiral wound membranes facilitated by marine algal bloom. Desalination. 2017; 424: 74-84. http://dx.doi.org/10.1016/j.desal.2017.09.035

[26] Farhat N.M, Javier L, Van Loosdrecht M.C.M, Kruithof J.C, Vrouwenvelder J.S. Role of feed water biodegradable substrate concentration on biofouling: biofilm characteristics, membrane performance and cleanability. Water Research. 2018. https://doi.org/10.1016/j.watres.2018.11.054

[27] Gao W, Liang H, Ma J, Han M, Chen Z. l, Han Z.s, Li G. Membrane fouling control in ultrafiltration technology for drinking water production: A review. Desalination. 2011; 272: 1-8. https://doi.org/10.1016/j.desal.2011.01.051

[28] Nagaraj V, Skillman L, Li D, Ho G. Review – Bacteria and their extracellular polymeric substances causing biofouling on seawater reverse osmosis desalination membranes. Journal of Environmental Management. 2018; 223: 586-599. DOI:10.1016/j.jenvman.2018.05.088

[29] Zhang M, Jiang S, Tanuwidjaja D, Voutchkov N, Hoek E.M. V, Cai B. Composition and variability of biofouling organisms in seawater reverse osmosis desalination plants. Applied and Environmental Microbiology. 2011; 77: 13: 4390-4398. DOI:10.1128/aem.00122-11

[30] Chun Y, Ha P.T, Powell L, Lee J, Kim D, Choi D, Lovitt R.W, Kim I.S, Mitra S.S, Chang I.S. Exploring microbial communities and differences of cartridge filters (CFs) and reverse osmosis (RO) membranes for seawater desalination processes. Desalination. 2012; 298: 85-92. https://doi.org/10.1016/j.desal.2012.05.007

[31] Shokralla Sh, Spall J.L, Gibson J.F, Hajibabaei M. Next-generation sequencing technologies for environmental DNA research. Molecular Ecology. 2012; 21: 8: 1794-1805. DOI:10.1111/j.1365-294x.2012.05538.x

[32] Bereschenko L.A, Heilig G.H.J, Nederlof M.M, Van Loosdrecht M.C.M, Stams A.J.M, Euverink G.J.W. Molecular characterization of the bacterial communities in the different compartments of a full-scale reverse-osmosis water purification plant. Applied and Environmental Microbiology. 2008; 74: 17: 5297-5304.: 10.1128/AEM.00387-08

[33] Horsch P, Gorenflo A, Fuder C, Deleage A, Frimmel F.H. Biofouling of ultra-and nanofiltration membranes for drinking water treatment characterized by fluorescence in situ hybridization (FISH). Desalination. 2005; 172 : 1: 41-52. https://doi.org/10.1016/j.desal.2004.05.009

[34] Wang Y. H, Wu Y.-H, Tong X, Yu T, Peng L, Bai Y, Zhao X. H, Huo Z, Ikuno N, Hu H.-Y. Chlorine disinfection significantly aggravated the biofouling of reverse osmosis membrane used for municipal wastewater reclamation. Water Research. 2019; 154: 246-257. DOI:10.1016/j.watres.2019.02.008

[35] Jiang L, Chen L, Zhu L. Fouling process of membrane distillation for seawater desalination: An especial focus on the thermal-effect and concentrating-effect during biofouling. Desalination. 2020; 485: 114457. https://doi.org/10.1016/j.desal.2020.114457

[36] Singh R, editors. Water and Membrane Treatment. Handbook of Membrane technology and engineering for water purification, 2nd ed. Elsevier Ltd; 2015. p. 81-177.

[37] Liberman B, Berman T. Analysis and monitoring: MSC — a biologically oriented approach. Filtration & Separation. 2006; 43: 4: 39-40. DOI:10.1016/s0015-1882(06)70869-2

[38] Manalo C. V, Ohno M, Okuda T, Nakai S, Nishijima W. Rapid novel test for the determination of biofouling potential on reverse osmosis membranes. Water Science & Technology. 2016; 73: 12. DOI: 10.2166/wst.2016.159

[39] Huang S, Voutchkov N, Jiang S.C. Investigation of environmental influences on membrane biofouling in a Southern California desalination pilot plant. Desalination. 2013; 319: 1-9. http://dx.doi.org/10.1016/j.desal.2013.03.016

[40] Dhakal N. Controlling biofouling in seawater reverse osmosis membrane systems [thesis]. UNESCO-IHE: Delft University of Technology: the Netherlands; 2017.

[41] Siebdrath N, Farhat N, Ding W, Kruithof J, Vrouwenvelder J. S. Impact of membrane biofouling in the sequential development of performance indicators: Feed channel pressure drop, permeability, and salt rejection. Journal of Membrane Science. 2019; 585: 199-207. DOI:10.1016/j.memsci.2019.05.043

[42] Al-Juboori R. A, Yusaf T, Aravinthan V. Investigating the efficiency of thermosonication for controlling biofouling in batch membrane systems. Desalination. 2012; 286: 349-357. DOI:10.1016/j.desal.2011.11.049

[43] Kurihara M, Takeuchi H, Ito Y. Review: a reliable seawater desalination system based on membrane technology and biotechnology considering reduction of the environmental impact. Environments. 2018; 5: 127. DOI:10.3390/environments5120127

[44] Filloux E, Wang J, Pidou M, Gernjak W, Yuan Z. Biofouling and scaling control of reverse osmosis membrane using one-step cleaning-potential of acidified nitrite solution as an agent. Journal of Membrane Science. 2015; 495: 276-283. DOI:10.1016/j.memsci.2015.08.034

[45] Wibisono Y, El Obied K.E, Cornelissen E.R, Kemperman A.J.B, Nijmeijer K. Biofouling removal in spiral-wound Nano filtration elements using two-phase flow cleaning. Journal of Membrane Science. 2014. http://dx.doi.org/10.1016/j.memsci.2014.10.016

[46] Vrouwenvelder J. S, Kruithof J. C, Van Loosdrecht M. C. M. Integrated approach for biofouling control. IWA Publishing. Water Science & Technology—WST. 2010; 62:11. DOI: 10.2166/wst.2010.747

[47] Beyer F, Laurinonyte J, Zwijnenburg A, Stams A. J. M, Plugge C. M. Membrane fouling and chemical cleaning in three full-scale reverse osmosis plants producing demineralized water. Journal of Engineering. 2017. https://doi.org/10.1155/2017/6356751

[48] Munshi H.A, Saeed M.O, Green T.N, Al-Hamza A. A, Farooque M. A, Ismail A. A. Impact of UV irradiation on controlling biofouling problems in NF-SWRO desalination process. In: International Desalination Association (IDA) World Congress Conference; 2005; Singapore.

[49] Bucs S. S, Farhat N, Kruithof J.C, Picioreanu C, Van Loosdrecht M.C.M, Vrouwenvelder J.S. Review on strategies for biofouling mitigation in spiral wound membrane systems. Desalination. 2018; 434: 189-197. https://doi.org/10.1016/j.desal.2018.01.023

[50] Bereschenko L.A, Prummel H, Euverink G.J.W, Stams A.J.M, Van Loosdrecht M.C.M. Effect of conventional chemical treatment on the microbial population in a biofouling layer of reverse osmosis systems. Water Research. 2011; 45; 2: 405-416. DOI:10.1016/j.watres.2010.07.058

[51] Scarascia G, Yap S. A, Kaksonen A. H, Hong P.-Y. Bacteriophage infectivity against *Pseudomonas aeruginosa* in saline conditions. Frontiers in Microbiology. 2018; 9:875. DOI: 10.3389/fmicb.2018.00875

[52] Yang W, Son M, Xiong B, Kumar M, Bucs S, Vrouwenvelder J.S, Logan B.E. Effective biofouling control using periodic H2O2 cleaning with CuO modified and polypropylene spacers. ACS Sustainable Chemistry & Engineering. 2019; 7: 9582-9587. DOI:10.1021/acssuschemeng.9b01086